I0073264

Electron Microscopy - Novel Microscopy Trends

*Edited by Masashi Arita
and Norihito Sakaguchi*

Published in London, United Kingdom

IntechOpen

Supporting open minds since 2005

Electron Microscopy - Novel Microscopy Trends
http://dx.doi.org/10.5772/intechopen.73375
Edited by Masashi Arita and Norihito Sakaguchi

Contributors
Ronald Gordon, Sei Saitoh, Jinfeng Yang, Haruo Sugi, Tsuyoshi Akimoto, Abeer Abd El Mohsen Abd El Samad Abd El Samad

© The Editor(s) and the Author(s) 2019
The rights of the editor(s) and the author(s) have been asserted in accordance with the Copyright, Designs and Patents Act 1988. All rights to the book as a whole are reserved by INTECHOPEN LIMITED. The book as a whole (compilation) cannot be reproduced, distributed or used for commercial or non-commercial purposes without INTECHOPEN LIMITED's written permission. Enquiries concerning the use of the book should be directed to INTECHOPEN LIMITED rights and permissions department (permissions@intechopen.com).
Violations are liable to prosecution under the governing Copyright Law.

(cc) BY

Individual chapters of this publication are distributed under the terms of the Creative Commons Attribution 3.0 Unported License which permits commercial use, distribution and reproduction of the individual chapters, provided the original author(s) and source publication are appropriately acknowledged. If so indicated, certain images may not be included under the Creative Commons license. In such cases users will need to obtain permission from the license holder to reproduce the material. More details and guidelines concerning content reuse and adaptation can be found at http://www.intechopen.com/copyright-policy.html.

Notice
Statements and opinions expressed in the chapters are these of the individual contributors and not necessarily those of the editors or publisher. No responsibility is accepted for the accuracy of information contained in the published chapters. The publisher assumes no responsibility for any damage or injury to persons or property arising out of the use of any materials, instructions, methods or ideas contained in the book.

First published in London, United Kingdom, 2019 by IntechOpen
IntechOpen is the global imprint of INTECHOPEN LIMITED, registered in England and Wales, registration number: 11086078, The Shard, 25th floor, 32 London Bridge Street
London, SE19SG – United Kingdom
Printed in Croatia

British Library Cataloguing-in-Publication Data
A catalogue record for this book is available from the British Library

Additional hard and PDF copies can be obtained from orders@intechopen.com

Electron Microscopy - Novel Microscopy Trends
Edited by Masashi Arita and Norihito Sakaguchi
p. cm.
Print ISBN 978-1-83881-882-1
Online ISBN 978-1-83881-883-8
eBook (PDF) ISBN 978-1-83881-884-5

We are IntechOpen,
the world's leading publisher of
Open Access books
Built by scientists, for scientists

4,300+

Open access books available

116,000+

International authors and editors

130M+

Downloads

Our authors are among the

151

Countries delivered to

Top 1%

most cited scientists

12.2%

Contributors from top 500 universities

CLARIVATE ANALYTICS
BOOK
CITATION
INDEX
INDEXED

WEB OF SCIENCE™

Selection of our books indexed in the Book Citation Index
in Web of Science™ Core Collection (BKCI)

Interested in publishing with us?
Contact book.department@intechopen.com

Numbers displayed above are based on latest data collected.
For more information visit www.intechopen.com

Meet the editors

Masashi Arita received his PhD degree in Solid State Physics from ETH Zurich, Switzerland, in 1987. He is now an associate professor at the Graduate School of Information Science and Technology, Hokkaido University, Sapporo, Japan. His major subject at present is in-situ electron microscopy of electronic devices.

Norihito Sakaguchi received his PhD degree in Materials Science from Hokkaido University, Japan, in 1999. He is now an associate professor at the Graduate School of Engineering, Hokkaido University, Sapporo, Japan. His major subject at present is nano-analysis using analytical transmission electron microscopy in metals and semiconductors.

Contents

Preface   XI

Chapter 1   1
Ultrafast Electron Microscopy with Relativistic Femtosecond
Electron Pulses
by Jinfeng Yang

Chapter 2   21
Electron Microscopic Recording of Myosin Head Power and
Recovery Strokes Using the Gas Environmental Chamber
by Haruo Sugi, Tsuyosi Akimoto and Shigeru Chaen

Chapter 3   35
Correlative Light-Electron Microscopy (CLEM) and 3D Volume
Imaging of Serial Block-Face Scanning Electron Microscopy (SBF-SEM)
of Langerhans Islets
by Sei Saitoh

Chapter 4   59
Transmission Electron Tomography: Intracellular Insight for the Future
of Medicine
by Abeer A. Abd El Samad

Chapter 5   69
Analytic Analyses of Human Tissues for the Presence of Asbestos and Talc
by Ronald E. Gordon

Preface

Since the inventions of the transmission electron microscope (TEM) and the scanning electron microscope (SEM), these instruments have greatly contributed to the progress of various research fields, such as crystallography, metallurgy, materials science and engineering, biology, medical science, etc. Intensive research and development of instrumentation, electron optical theory, and methodology have been performed by many researchers and technicians, and imaging quality has been improved. In the last few decades, because of developments in field emission (FE) electron sources and aberration correctors, highly resolved imaging is now widely available by using up-to-date electron microscopes such as the aberration-corrected TEM/STEM and FE-SEM. In addition, analytical imaging has become possible in less time because of the development of detectors and operation software.

In this tide of microscopy, various microscopic methods have also been developed to make clear many unsolved problems. For example, high-speed investigations are possible using a pulse beam TEM and environmental microscopy enables observations in the atmosphere. Other examples include electron microscopy combined with other measurement systems like in-situ TEM and correlative microscopy, and tomography giving 3D information of engineering materials and biological tissues. In addition, development of methodologies in sample preparation is also an important issue to achieve results reflecting original sample characteristics.

In this book, we have collected a number of examples concerning these subjects. In Chapter 1, a relativistic ultrafast electron microscope with femtosecond electron pulses and ultrahigh voltage microscopy is reviewed. In addition to details of instrumentation, observation results to evaluate its performance are shown. In Chapter 2, an example of environmental microscopy applied to biology is reviewed. Using a specially designed system, clear movement of a myosin head under different conditions in a gas atmosphere is demonstrated. In Chapter 3, correlative light-electron microscopy and serial block face scanning electron microscopy are reviewed. Following a general introduction of these methods and detailed experimental procedures, results of human and mouse pancreases are demonstrated. In Chapter 4, applications of high-voltage transmission electron tomography to biomedical investigations are reviewed. Various examples to investigate cell organelles are shown. In Chapter 5, TEM applications in environmental medicine are reviewed. Specified topics here are the diseases caused by asbestos and talc fibers. Procedures and protocols of the sampling of these powders contained in human tissues without contaminants are discussed.

Although this book includes a limited number of topics, we think that the content in each chapter will be impressive to the reader. We hope this book will contribute to future advances in electron microscopy, materials science, biomedicine as well as other research fields.

We are grateful to all authors for their contributions to this book and their efforts to complete the chapters. We also acknowledge the IntechOpen publishing team, especially Martina Brkljačić and Markus Mattila for cooperation in the publishing process.

Dr. Masashi Arita
Graduate School of Information Science and Technology,
Hokkaido University,
Japan

Dr. Norihito Sakaguchi
Faculty of Engineering,
Hokkaido University,
Japan

Chapter 1

Ultrafast Electron Microscopy with Relativistic Femtosecond Electron Pulses

Jinfeng Yang

Abstract

An ultrafast electron microscope (UEM) with a femtosecond temporal resolution is a "dream machine" desired for studies of ultrafast structural dynamics in materials. In this chapter, we present a brief overview of the historical development of current UEMs with nonrelativistic electron pulses to illustrate the need for relativistic-energy electron pulses. We then describe the concept and development of a UEM with relativistic femtosecond electron pulses generated by a radio frequency (RF) acceleration-based photoemission gun. The technique of RF electron gun and physical characteristics of the relativistic electron pulses are described. Demonstrations of UEM images acquired using approximately 100 fs long electron pulses with energies of 3 MeV are presented. Finally, we report a single-shot diffraction imaging methodology in the UEM with a relativistic femtosecond electron pulse for studies of laser-induced ultrafast phenomena in crystalline materials.

Keywords: ultrafast electron microscopy, ultrafast electron diffraction, femtosecond electron pulse, relativistic electron beam, structural dynamics, radio-frequency electron gun

1. Introduction

It is well-known that transmission electron microscope (TEM) is one of the most powerful imaging instruments. Currently, the TEM enables imaging of three-dimensional (3D) structures on the atomic scale. However, the temporal resolution of TEM is limited by the video-camera recording rate (millisecond), as continuous electron beams produced by DC-acceleration-based thermionic or field-emission sources are used in conventional TEMs. In order to overcome this millisecond temporal resolution limit in TEM, in 1987, Bostanjoglo et al. [1, 2] proposed a technique with a nanosecond-triggered beam-blanking unit in TEM to produce electron beam flashes and succeeded in observation of laser-induced phase transitions in amorphous Ge films with a nanosecond resolution. Two years later, they developed time-resolved electron microscopy with a laser-triggered photoemission electron gun [3, 4] and paved the way for studies and imaging of fast dynamic events with pulsed electron beams. In the time-resolved electron microscopy (pump-probe technique), an initialing pump laser pulse excites the materials (sample) to produce a transient state, and then a later analytical probe electron pulse detects it at a correlated time with the pump pulse, as shown in **Figure 1**. It is worth noting that the temporal resolution is determined by the durations of the pump laser pulse and probe electron pulse. It is

IntechOpen

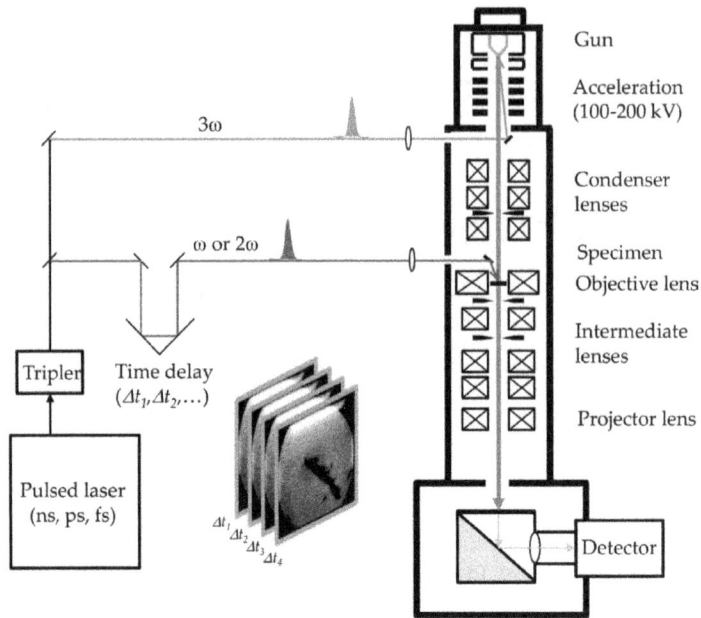

Figure 1.
General schematic of time-resolved electron microscopy using electron pulses.

not limited by the speed of the electron detector. Therefore, the generation of short electron pulses is very significant.

In the past decade, both temporal and spatial resolutions of the time-resolved electron microscopy were improved by several research groups [5–15] by developing high-brightness electron sources and optimizing TEM electron optics. Currently, there are two main approaches to time-resolved electron microscopy: (1) single-shot approach, which uses high-charge electron pulses and (2) stroboscopic approach, which uses "single" electron pulses. The first approach, pioneered by Bostanjoglo et al. (described above), aims to generate a sufficient number of electrons in the pulse to form an image with a single shot. The main advantage of this approach is that the studied structural dynamics or processes do not need to be perfectly reversible. This approach has been developed later at the Lawrence Livermore National Laboratory (LLNL) [5, 6] using a modified JEOL 2000FX for dynamic transmission electron microscopy (DTEM). Single-shot imaging with spatial and temporal resolutions of 10 nm and 15 ns, respectively, has been achieved using a 15-ns electron pulse containing 5×10^7 electrons. DTEM has achieved a spatial resolution of approximately 5 nm and has been applied to various projects, including laser-induced melting of metals, reactions in solid states, and crystallization of amorphous semiconducting materials. However, a serious problem in this approach is the significant space-charge effect due to the Coulomb repulsion of electrons. The high current of the pulses limits the overall temporal and spatial resolutions of the instrument, even with an optimized microscope source, good column, and excellent detector.

The stroboscopic approach uses a femtosecond laser source to generate an extremely low-current electron pulses ("single" electron pulses) at the sample to maintain high temporal and spatial resolutions. A key factor in the stroboscopic approach is to have only a single electron in the column at any time to reduce space-charge effects. Images are constructed from ~10^7 of these single-electron pulses. The first single-electron imaging was demonstrated by Zewail et al. [7–9] at the California Institute of Technology (Caltech) using a modified 120-kV TEM as a first-generation ultrafast electron microscope (UEM). Later, they constructed a second-generation UEM-2, using a hybrid 200-kV TEM, and achieved spatial and temporal resolutions of 3.4 Å and 250 fs, respectively [8, 9]. The temporal resolution Δt (root-mean-square (RMS)) in UEM is determined by the laser pulse width Δt_{hv} and temporal broadening Δt_{KE} due to excess energy of electrons emitted from the photocathode. It can be calculated as [9]

$$\Delta t = \sqrt{\Delta t_{hv}^2 + \Delta t_{KE}^2}, \tag{1}$$

$$\Delta t_{KE} = \frac{d}{eV}\sqrt{\frac{m_0}{2}}\frac{\Delta E_i}{\sqrt{E_i}}, \tag{2}$$

where e is the electron charge, m_0 is the electron rest mass, ΔE_i is the energy spread of electrons emitted from the cathode, E_i is the mean energy, and V is the potential across the distance d between the cathode and anode. $\Delta t_{KE} \sim 250$ fs at $\Delta E_i = 0.1$ eV and 650 fs at $\Delta E_i = 0.5$ eV [9]. Recently, there are many research activities [10–15] focused on improving the electron source and electron optics inside the microscope to achieve better temporal and spatial resolutions. However, the current UEM has restrictions: (1) the specimen must be pumped ~10^7 times by the laser, which implies that the studied processes must be perfectly reversible, that is, the sample must heal between pulses; (2) the interval between two pulses needs to be longer than the recovery time of the specimen, thus limiting the repetition frequency of laser excitation; and (3) instrumental instabilities such as specimen drift are inevitable at long exposure times. Therefore, for most experiments under realistic conditions, it is necessary to operate with a larger number of electrons per pulse.

All of the current DTEMs and UEMs use a DC-acceleration-based photoemission electron gun to generate electron pulses at energies of 100–200 keV. For such low-energy electron pulses (nonrelativistic electron pulses), the space-charge effect due to the Coulomb repulsion of electrons is a serious problem. It not only broadens the pulse duration but also increases the energy spread and beam divergence during the acceleration (particularly near the cathode) and propagation to the specimen. We can reduce the space-charge broadening in the propagation with a high-energy electron beam, that is, with the use of ultrahigh-voltage electron microscopy. However, the space-charge effect near the cathode can be reduced only by instantaneously accelerating the electrons to high energies with a high electric field. This is very difficult or impossible, as the maximum static electric field in the DC gun is determined by the vacuum breakdown limit of ~10 MV/m. The current state-of-the-art DC guns have been developed to generate 600-fs electron pulses with up to 10,000 electrons per pulse [16]. An approach of electron pulse compression based on the radio frequency (RF) technology was proposed to control the space-charge effect and generate high-current femtosecond electron pulses at the sample position [17–20]. However, this approach is not suitable for microscopy, because of the relatively large energy spread of the electron pulses. From the theoretical modeling and experimental study, it is obvious that it is impossible to use the DC electron gun parameters to attain femtosecond or picosecond electron pulses at the sample position with a reasonable electron number of 10^7 (or larger) for the single-shot imaging.

Recently, an advanced accelerator technology of RF-acceleration-based photoemission electron guns has been successfully applied for ultrafast electron diffraction (UED) with a single-shot measurement [21–30]. The RF gun is usually operated with a high RF electric field equal to or higher than 100 MV/m, which is 10 times higher than those of the DC guns. Therefore, the electrons emitted from the photocathode can be quickly accelerated into the relativistic-energy region to minimize the space-charge effects in the pulse, yielding a femtosecond or pico-second pulse containing a large number of electrons. Yang et al. have developed the first prototype of relativistic UEM using the RF gun technology at the Osaka University, have succeeded in generation of high-brightness relativistic electron pulses with a pulse duration of 100 fs containing 10^7 electrons at an energy of 3 MeV [31, 32], and have demonstrated the first single-shot imaging using such femtosecond electron pulses [33–36]. This promising approach has a large potential to study ultrafast dynamics in materials, as there are several crucial advantages over nonrelativistic UEM systems:

- The relativistic energy reduces the space-charge effect, which consequently maintains the temporal confinement and beam brightness during the propaga-tion to the sample. It is possible to perform single-shot imaging with femto-second temporal resolution to observe reversible and irreversible processes in materials.

- Relativistic-energy electrons significantly enhance the extinction distance for elastic scattering and provide structural information, essentially free from multiple scattering and inelastic effects [37]. Our previous UED study of struc-tural dynamics of laser-irradiated gold nanofilms indicates that the kinematic theory can be applied in the case of 3-MeV probe electrons with the assump-tion of single scattering events [29, 30]. This enables to easily understand and explain structural dynamics.

- A thick sample can be used for measurement, thereby obviating the require-ment to prepare suitable thin samples. The utilization of the relativistic electron pulse overcomes the loss of temporal resolution due to the velocity mismatch in samples. Furthermore, the relativistic UEM is suitable for in situ observations as there are large area and space inside the objective lens to install various specimens.

In this chapter, we introduce a UEM with relativistic femtosecond electron pulses, including information of an RF photoemission gun, concept of relativistic UEM, and demonstration experiments with relativistic femtosecond electron pulses.

2. Relativistic UEM

The relativistic UEM consists of a 1.6-cell S-band (2.856 GHz) photocathode RF gun and microscopy column including an electron illumination system, objective lens, and imaging system. **Figure 2** shows a prototype of relativistic UEM, which has a height of 3 m and diameter of 0.7 m.

2.1 Photocathode RF gun

Photocathode RF gun is a high-brightness electron source in the particle accel-erator field, which has been widely applied in free-electron lasers [38, 39], and is considered for use in the next linear colliders [40, 41]. The RF gun used in our relativistic UEM consists of two RF cavities: a half cell and full cell. The accelerating RF is 2.856 GHz, belonging to the so-called S-band corresponding to a wavelength of λ = 104.969 mm. The length of the full-cell cavity is $\lambda/2$ = 52.4845 mm, while the length of the half-cell is 31.49 mm, equal to 0.6 times $\lambda/2$, as numerical studies

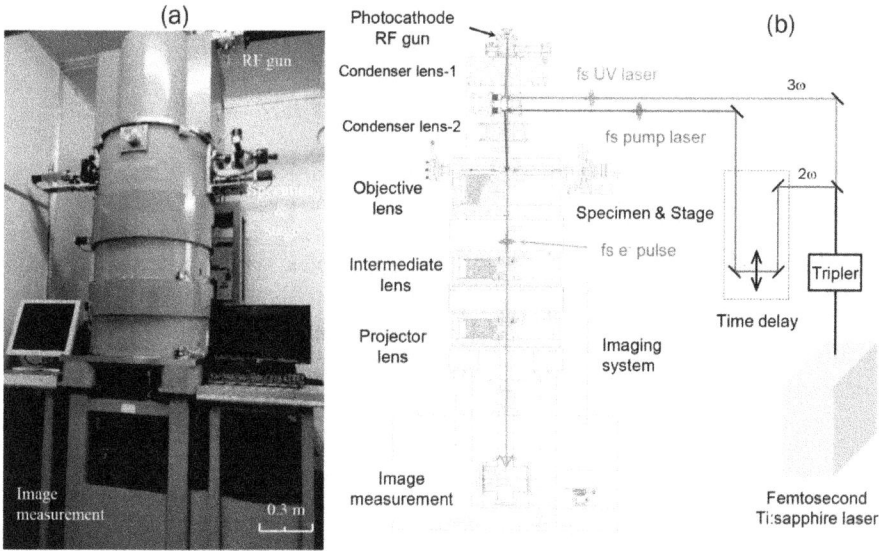

Figure 2.
(a) Photograph and (b) conceptual design of a relativistic UEM with femtosecond electron pulses.

show that more optimal performances are obtained if the half-cell cavity length is 0.6 (rather than 0.5) times the full-cell length. The cross section of the actual RF cavities is shown in **Figure 3(b)**. The shapes of the cavities are optimized to reduce both RF-induced emittance and energy spread. The cavities and other components are carefully fabricated and brazed together to minimize the dark current due to the field emission. A very fine copper photocathode is located at a position of a high-accelerating RF field in the half-cell cavity and is illuminated by the third harmonic of a Ti:sapphire laser (ultraviolet (UV) emission: 266 nm, pulse duration: 100 fs). When the electrons leave the cathode, they are accelerated by a negative electric field in the cavity, as shown in **Figure 3(a)**.

Generally, the RF gun uses the TM_{010} transverse magnetic mode with a phase shift of π between the half cell and full cell. The linear components of the electric and magnetic fields in the RF cavities can be assumed to be [42, 43]

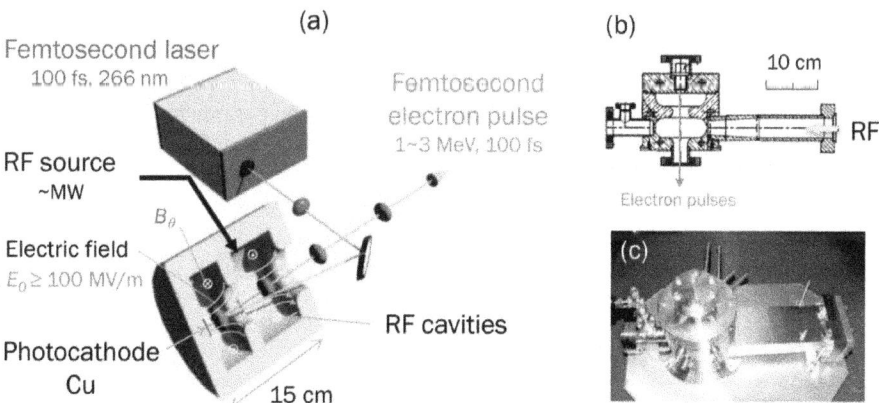

Figure 3.
(a) Schematic of femtosecond electron pulse generation in the photocathode RF gun, (b) the cross section of the RF cavities, and (c) photograph of the RF gun.

$$E_z = E_0 \cos kz \sin(\omega t + \phi_0) \text{ at } r = 0,$$

$$E_r = \frac{kr}{2} E_0 \sin kz \sin(\omega t + \phi_0), \tag{3}$$

$$B_\theta = c\frac{kr}{2} E_0 \cos kz \cos(\omega t + \phi_0),$$

where $E_0 \geq 100$ MV/m is the peak accelerating field, $k = 2\pi/\lambda$, $\omega = ck$, c is the velocity of light, and ϕ_0 is the initial RF phase when the electron leaves the cathode surface ($z = 0$) at $t = 0$. For short electron pulses with Gaussian distributions in longitudinal and transverse directions, Kim [42] and Travier [44] reported analytical expressions for the following parameters at the gun exit in practical units: electron energy E in MeV, RMS relative energy spread $\Delta E/E$, RMS angular divergence σ'_x in mrad, normalized RMS RF-induced emittance ε_{rf}, and normalized RMS space-charge-induced emittance ε_{sc} in mm-mrad:

$$E = 75(n + 0.5)\frac{E_0}{f}, \tag{4}$$

$$\frac{\Delta E}{E} = 2 \times 10^{-6}\frac{f\sigma_b}{n + 0.5}, \tag{5}$$

$$\sigma'_x = 0.511\frac{E_0\sigma_x}{E + 0.511}, \tag{6}$$

$$\varepsilon_{rf} = 2.73 \times 10^{-11} E_0 f^2 \sigma_x^2 \sigma_b^2, \tag{7}$$

$$\varepsilon_{sc} = 3.76 \times 10^3 \frac{Q}{E_0(2\sigma_x + \sigma_b)}, \tag{8}$$

where E_0 is the peak accelerating field in MV/m, f is the RF in MHz, n is the number of full-cell cavities, Q is the electron charge of the pulse in nC, and σ_b and σ_x are the RMS pulse duration in ps and RMS transverse beam size in mm, respectively, which are dependent on the laser pulse width and spot size on the cathode. These expressions are useful to estimate the beam parameters for the consideration and design of a new RF gun in other instruments. For example, for a 1.5-cell S-band RF gun at $E_0 = 100$ MV/m with a 100-fs and 0.1-mm laser, we estimate the following beam parameters: $E = 3.9$ MeV, $\Delta E/E = 3.8 \times 10^{-4}$, $\sigma'_x = 1.1$ mrad, $\varepsilon_{rf} = 2.2 \times 10^{-6}$ mm-mrad, and $\varepsilon_{sc} = 1.3 \times 10^{-2}$ mm-mrad at $Q = 0.1$ pC. These estimated parameters are in agreement with a particle simulation with the space-charge effect, as described below.

For a short electron pulse with a small beam size, the RF-induced emittance is negligible in the RF gun. In this case, the thermal emittance (initial emittance) at the cathode is dominant. Assuming an isotropic emission into a half sphere in front of the cathode surface, the thermal emittance can be expressed in terms of the RMS incident laser spot size on the cathode σ_r and a kinetic energy of the emitted electrons E_{kin}:

$$\varepsilon_{th} = \sigma_r\sqrt{\frac{2E_{kin}}{3m_0c^2}}, \tag{9}$$

$$E_{kin} = h\upsilon - \varphi + \varphi_{schottky}, \tag{10}$$

$$\varphi_{schottky} = \alpha\sqrt{\beta E_0}\sin\varphi_0, \tag{11}$$

where $h\nu$ s the laser photon energy, φ is the work function of the cathode, $\phi_{schottky}$ is the reduction in the potential wall barrier due to the Schottky effect, and α and β are a constant and field-enhancement parameter, respectively, which are determined by the roughness and clearness of the cathode surface. Finally, the total beam emittance is

$$\varepsilon = \sqrt{\varepsilon_{rf}^2 + \varepsilon_{sc}^2 + \varepsilon_{th}^2}. \tag{12}$$

For a very fine copper photocathode with a 266-nm laser under the conditions of E_0 = 100 MV/m and ϕ_0 = 30°, the measured thermal emittance exhibited a linear relationship with the RMS laser spot size: ε_{th} [mm-mrad] = 0.74 × σ_r [mm] [45], yielding an E_{kin} = 0.42 eV.

In electron microscopy, the electron gun is required to generate a low-emittance and low-energy-spread beam (normalized emittance ≤0.1 mm-mrad, energy spread ≤10^{-4}). The beam dynamics in the RF gun can be calculated by a theoretical simulation using the General Particle Tracer (GPT) code [46]. **Figure 4** presents the simulation results of relativistic femtosecond-pulsed electron beams at the specimen position in the UEM generated from a 1.6-cell S-band RF gun. The results of the beam dynamics show that (1) the minimum energy spread is obtained at the launch phase of approximately 20°. At ϕ_0 ≥ 60°, the energies of the electrons decrease leading to the large increases in the energy spread due to the space-charge effects during the electron propagation. The slight increase in the energy spread

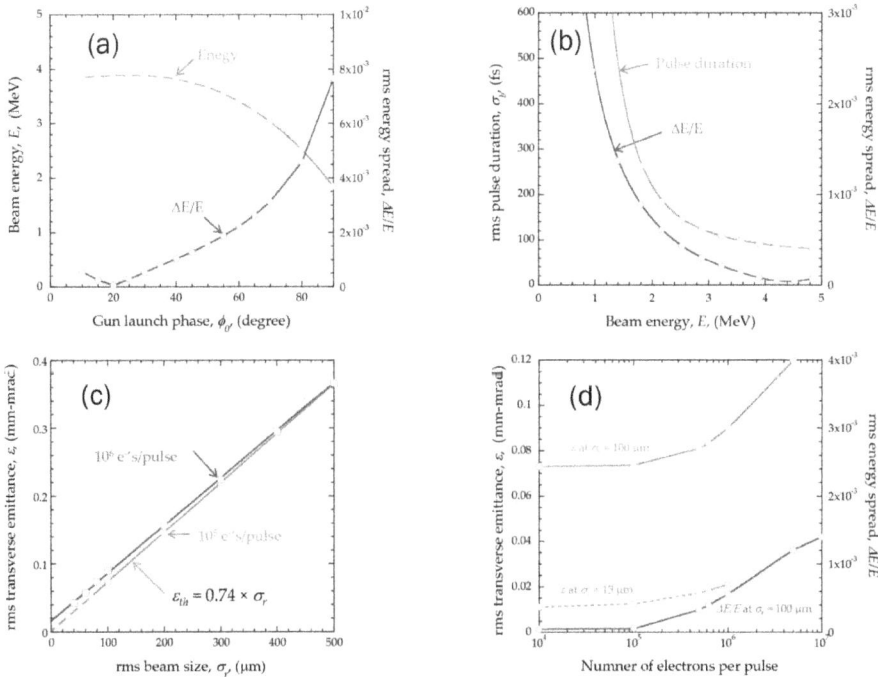

Figure 4.
Simulation results of the beam energy, pulse duration, emittance, and energy spread of electron pulses generated by the RF gun driven with a 100-fs laser under the conditions of (a) Q = 0.16 pC and E_o = 100 MV/m, (b) Q = 0.16 pC and ϕ_0 = 20°, (c) E = 3.8 MeV and ϕ_0 = 20°, and (d) E = 3.8 MeV and ϕ_0 = 20°.

at $\phi_0 < 20°$ is caused by the space-charge effect near the cathode as the actual RF electric field is decreased by $E_0\sin\phi_0$, (2) large increases in both energy spread and pulse duration are observed at an electron energy smaller than 3 MeV, and (3) the transverse emittance of the femtosecond electron pulses generated from the RF gun is dominated mainly by the thermal emittance if the number of electrons in the pulse is 10^6 (or smaller). This allows us to generate low-emittance electron pulses by focusing the laser spot on the cathode, as described in Eq. (9). The increase in the energy spread due to the space-charge effect is also negligible at an electron number smaller than or equal to 10^6 per pulse. The theoretical modeling and particle simulation indicate that the RF gun could generate high-current femtosecond electron pulses with excellent characteristics, including a normalized RMS transverse emittance of 0.02 mm-mrad, energy spread of 10^{-4} to 10^{-5}, and pulse duration of 100 fs with 10^6 electrons per pulse at an energy larger than or equal to 3 MeV. The peak brightness of such electron pulses B_p can be calculated by

$$B_p = (\beta\gamma)^2 \frac{Q}{\varepsilon^2 \sigma_b}, \qquad (13)$$

where $\beta = v/c$, v is the electron velocity, and γ is the normalized relativistic energy. From the given parameters, we can calculate the peak brightness, $B_p = 2 \times 10^{17}$ A/m²·sr. Recently, the development of a high-repetition-rate normal-conducting RF gun at 1000 Hz was proposed in our research group. In the near future using this RF gun, the time-averaged brightness of the femtosecond-pulsed electron beam can be potentially increased to $B = 2 \times 10^7$ A/m²·sr, corresponding to the brightness of the 100-kV thermionic emission gun used in modern TEMs. In addition, if we can further reduce the energy spread, the RF gun will pave the way for generation of a practical high-brightness femtosecond-pulsed electron beam for UEM.

2.2 UEM column

2.2.1 Electron illumination system

The electron illumination system comprises two condenser lenses and condenser aperture to control and transfer the electron pulses from the RF gun on the specimen, as shown in **Figure 5**. An aperture with three pinholes with diameters of 0.5, 1, and 2 mm is made of a 1-mm-thick molybdenum metal and is installed between the two condenser lenses. The distance from the photocathode to the specimen is 0.8 m. However, there are two significant differences from the normal TEMs: (1) there is no gun crossover in the RF gun and (2) we cannot produce a crossover between the two condenser lenses owing to the space-charge effect (i.e., the condenser lenses do not condense). Therefore, in this illumination system, the image of the beam spot on the cathode acts as the object plane of the first condenser lens as the beam size is minimum in the RF gun. It is worth noting that a virtual source image behind the cathode may be better for the object plane of the first lens as the source size is smaller than that on the cathode, as shown in **Figure 5**. We adjust the first lens to create a parallel electron beam. The aperture stops the large-divergence electrons to further reduce the emittance, yielding a small illumination convergence angle at the specimen. Finally, we used the second lens to create a parallel beam or convergent beam on the specimen. The parallel beam is used primarily for UEM imaging and selected-area diffraction (SAD), while the convergent beam is used mainly for convergent-beam electron diffraction (CBED). In the experiments, a small convergence angle of $\alpha = 31$ μrad was achieved with the condenser aperture with the pinhole diameter of 0.5 mm in the parallel-beam operation mode.

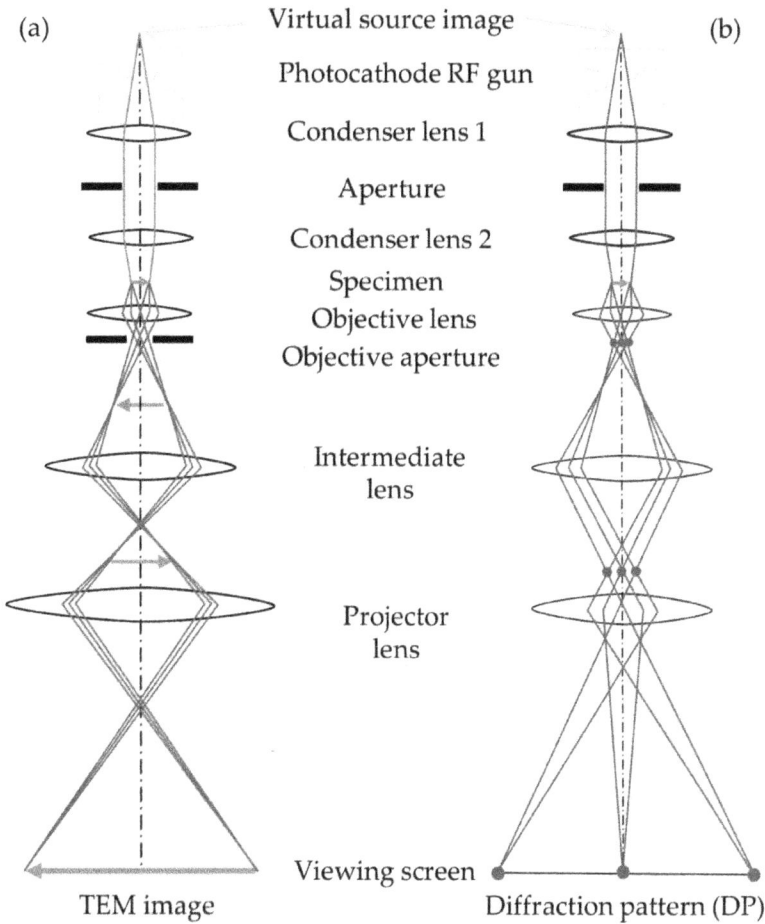

Figure 5.
Two basic operations of the UEM imaging system: (a) imaging mode: Projection of a TEM image onto the viewing screen and (b) diffraction mode: Projection of diffraction patterns (DPs) onto the screen.

2.2.2 Objective lens

The objective lens is the most important lens in a TEM. It forms TEM images or DPs, magnified by the other lenses. It is very challenging to construct this lens as the specimen must be located close to its center. In the UEM, two asymmetrical and separable pole pieces (upper and lower) are used for the objective lens. The gap between the pole pieces is 19 mm. It allows us to insert both specimen and aperture between the pole pieces. With the large gap, it is straightforward to design specimen holders for various tasks such as tilting, rotation, heating, cooling, straining, etc. The gap-to-bore ratio is 1.46 for creation of a strong magnetic field and aberration suppression. We carefully designed and produced all of the pole pieces, iron circuits, coils, and water-cooling components. The pole pieces were made of a soft magnetic alloy (Permendur) with an iron-to-cobalt content ratio of 1:1. The saturation magnetic flux density is 2.4 T. The objective lens is a very strong lens. The maximum magnetic field strength at the center of the pole pieces is 2.3 T under the magnetomotive force of 35 kA·turns. The focal length is 5.8 mm for a 3-MeV electron beam. The diameter of the lens is 0.7 m, as shown in **Figure 6(a)**.

The specimen is placed close to the center of the pole pieces with a side-entry method and manipulated by a five-axis motorized stage. The sample can be pumped by a pulsed laser beam, as shown in **Figure 2**. An objective aperture with a pinhole diameter of 0.3 mm is inserted at the back-focal plane of the objective lens to block scattered electrons in the imaging operation mode.

2.2.3 Imaging system

The imaging system uses two magnetic lenses (intermediate and projector lenses) to magnify the TEM image or DP produced by the objective lens and to project it onto a viewing screen (scintillator) through a charge-coupled device (CCD) camera. In the intermediate and projector lenses, two symmetrical and separable pole pieces are used. The gap is 20.8 mm, while the gap-to-bore ratio is ~1.6 for minimization of the aberration coefficients. The pole pieces in the intermediate and projector lenses are made of Permendur and pure iron, respectively. The distance between the objective and intermediate lenses is 0.6 m. The distance between the intermediate and projector lenses is ~0.4 m. The maximum magnetic field strengths are 2.3 T in the intermediate lens and 1.2 T in the projector lens.

In a time-resolved image measurement with the pump-probe technique, the spread of the detector (recording rate of the video-camera) does not limit the temporal resolution; however, a high-sensitivity detection of electron waves is crucial. In particular, acquisition of images in a single shot is necessary in order to observe irreversible processes. In addition, as mentioned above, the space-charge effect causes increases in the pulse duration, emittance, and energy spread, even in relativistic electron beams. Low-current electron pulses should be used in the UEM. Therefore, the image recording system should be highly sensitive to every electron. In the UEM, in order to achieve a high sensitivity to MeV electron detection with a high damage threshold, we chose a Tl-doped CsI columnar crystal scintillator equipped with a fiber optic plate (Hamamatsu Photonics) to convert the relativistic-energy TEM images or DPs into optical images. The light generated by the scintillator is propagated by a reflective mirror (at 45°) consisting of aluminum deposited on a 5-μm-thick polymer, and finally the optical images are detected with an electron-multiplying CCD (EMCCD) with 1024 × 1024 pixels, as shown in **Figure 6(b)**. The effective detection area of the scintillator is 50 × 50 mm^2, while the distance from the specimen to the scintillator is 1.8 m. The sensitivity of the whole detection system is 3×10^{-3} counts/electron.

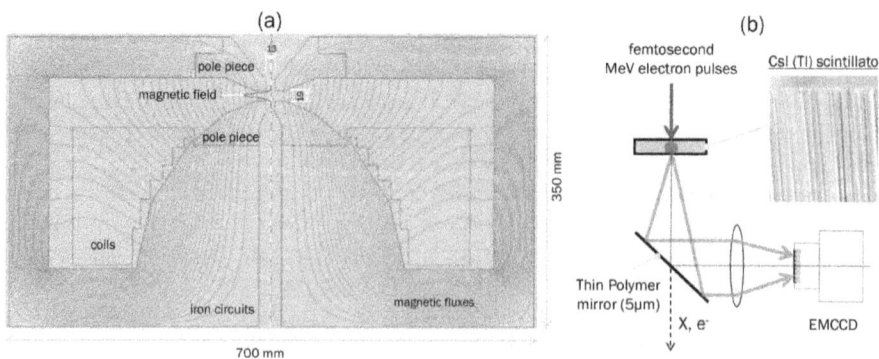

Figure 6.
(a) Cross section of the objective lens and (b) image detection system with a CsI (Tl) scintillator.

2.3 Femtosecond laser system

A femtosecond Ti:sapphire laser (Spectra-Physics) is used to generate femto-second electron beam pulses and to excite the specimen. The laser consists of a Ti:sapphire laser oscillator with a pulse width of 90 fs (Tsunami, central wave-length: 800 nm), regenerative amplifier including a pulse compressor, and a wave-length converter. The femtosecond laser oscillator is synchronized to a 79.3-MHz RF signal, corresponding to 1/36 of the 2856-MHz RF used for the electron generation in the RF gun, with a time-to-lock piezoelectric device. The time jitter between the laser pulse and RF phase is 61 fs [36]. The laser pulses generated from the oscillator are fed to the regenerative amplifier and amplified to ~1 mJ per pulse. The regenera-tive amplifier is driven by a green laser with a highly stable repetition rate of 1 kHz (Empower, wavelength: 532 nm, output: 10 W).

The amplified femtosecond pulses are converted to the third harmonics by a wavelength converter (Tripler) composed of two nonlinear crystals (SHG and THG) and time plate for pulse delay adjustment. The third-harmonic pulses (UV wavelength: 266 nm) with the maximum energy of 70 µJ per pulse are illuminated onto the copper cathode to generate femtosecond electron pulses. The residual fundamental (wavelength: 800 nm) and second-harmonic (wavelength: 400 nm) pulses are used to pump the samples. The time delay between the pump laser pulse and probe electron pulse is changed with an optical delay located on the pump laser beam line for time-resolved experiments. Both laser beams used for the electron generation and sample excitation can be focused to a spot size with a full width at half maximum (FWHM) smaller than 30 µm at the corresponding target.

3. Observations with relativistic femtosecond electron pulses

3.1 UEM (imaging mode)

In the imaging operation mode of the relativistic UEM, we adjust the inter-mediate lens so that its object plane is the image plane of the objective lens and then project the image onto the scintillator with the projection lens, as shown in **Figure 5(a)**. The objective aperture with a pinhole diameter of 0.3 mm inserted at the back-focal plane of the objective lens is used to block the diffraction pat-terns and other scattered electrons.

Figure 7 shows bright-field UEM images of polystyrene latex particles, with diameters of 1.09 µm and gold nanoparticles with diameters of 400 ± 20 nm, dispersed on a carbon film pasted on a copper mesh, observed with a pulse integra-tion measurement [35]. The magnification of the images is approximately 1500×. The energy of the electron pulses used for the observations is 3.1 MeV. The pulse duration is approximately 100 fs. In this experiment, a condenser aperture with a pinhole diameter of 0.5 mm is used to collimate the electron beam. The repetition rate of the electron pulses is 10 Hz. The normalized RMS emittance and electron charge in the pulses are 0.12 mm-mrad and 1 pC at the specimen position, respec-tively. From these parameters, the estimated peak brightness of the pulses in the observations is $B_p = 3.5 \times 10^{16}$ A/m^2·sr. For these electron pulses, acceptable images of both polystyrene latex particles and gold nanoparticles can be obtained using our relativistic UEM with 2000- and 500-pulse integrations, respectively.

At a low-magnification observation condition, the single-shot observation of UEM images with a relativistic femtosecond electron pulse is achievable. **Figure 8** presents UEM images of single-crystal gold obtained with a single shot and 10- and

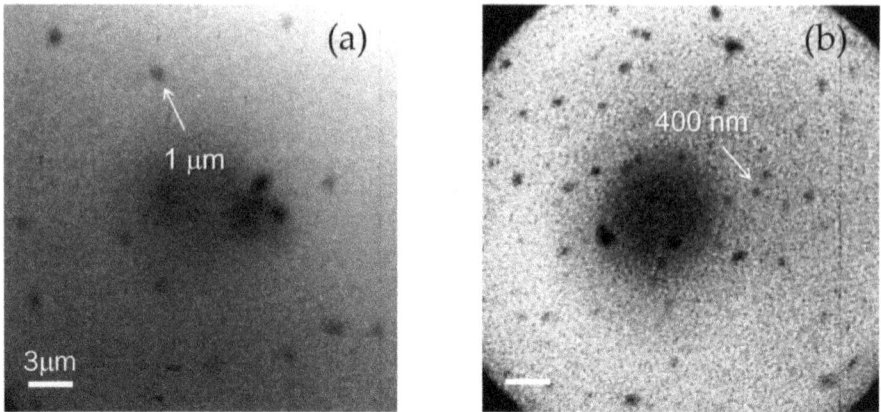

Figure 7.
UEM images (bright field) of (a) polystyrene latex particles with a diameter of 1.09 μm and (b) gold nanoparticles with a diameter of 400 ± 20 nm obtained by 2000 and 500-pulse integrations, respectively, with 100-fs electron pulses at an energy of 3.1 MeV [35].

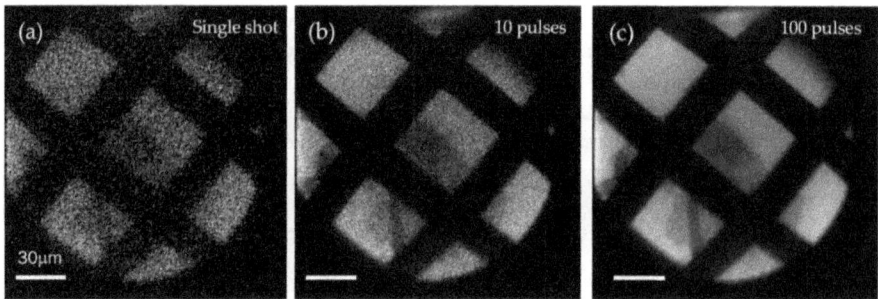

Figure 8.
UEM images (bright field) of gold single-crystals placed on a gold mesh observed by (a) a single shot and (b) 10- and (c) 100-pulse integrations with 100-fs electron pulses at an energy of 3.1 MeV.

100-pulse integrations. The sample is a single-crystalline gold film with a thickness of 10 nm placed on a gold mesh. Although the signal-to-noise ratio is still not sufficient in the single shot, an acceptable image can be obtained with 10 pulses (or more), as shown in **Figure 8(b)** and **(c)**. This demonstrates that the single-shot observation of UEM images with a relativistic electron pulse is achievable if the brightness of the pulses can be increased by one order of magnitude, $B_p \sim 10^{17}$ A/m^2·sr, which can be realized, as described in Section 2.1.

3.2 UED (diffraction mode)

In the relativistic-energy electron diffraction measurement (diffraction mode), we adjust the imaging-system lenses so that the back-focal plane of the objective lens acts, as the object plane for the intermediate lens, and then use the projection lens to project the DPs onto the scintillator, as shown in **Figure 5(b)**. It is worth noting that in the diffraction mode, if the electron beam is too intense, it will damage the viewing screen or saturate the CCD camera. Therefore, we could generate a smaller beam to reduce the illuminated area of the specimen contributing to the DP on the screen. The use of a selected-area aperture inserted at the image plane of the objective lens is a better choice to create a virtual aperture in the plane of the specimen, thus yielding only the SAD pattern. In the diffraction mode, the objective aperture is not used.

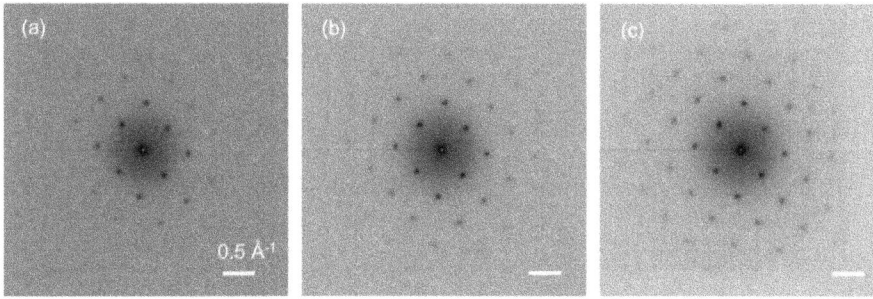

Figure 9.
DPs of a (100)-oriented single-crystalline gold measured with (a) a single shot and (b) 10- and (c) 100-pulse integrations. The energy of the electron pulses is 3.1 MeV, while the number of electrons in each pulse is 6.3 × 10⁶, corresponding to 1 pC per pulse.

In order to check the contrast of the DPs and resolution, we generate a parallel electron beam with a diameter of 0.5 mm to illuminate the specimen using a condenser aperture with a diameter of 0.5 mm and weakened second condenser lens in the illumination system. The characteristics of the electron pulses are the same as those in the imaging operation mode. **Figure 9** presents the DPs of the single-crystal gold observed by a single shot and 10- and 100-pulse integrations. The sample is a single-crystalline gold film with a thickness of 10 nm placed on a gold mesh. The energy of the electron pulses is 3.1 MeV. The electron charge per pulse is 1 pC. Data in **Figure 9** show that (1) sharp DPs and high contrast can be obtained with the relativistic femtosecond electron pulses, (2) higher order spots of (−420) and (4−20) from the single-crystal gold with scattering vectors up to 1.1 Å^{-1} can be clearly captured with a single shot, and (3) the RMS width of the zeroth-order spot (000) in the single shot is measured to be 0.018 Å^{-1}, indicating an excellent spatial resolution for the relativistic diffracted beam.

Based on the width of the (000) spot and measured distance of the diffraction spots from the (000) position, the estimated RMS illumination convergence angle α of the electron beam at the specimen is α = 31 μrad. This convergence angle is two orders of magnitude smaller than that of nonrelativistic UED. Additionally, the coherence of the electron source is an important parameter in diffraction imaging, particularly in terms of spatial coherence (transverse coherence), which determines the sharpness of the DPs and diffraction contrast in the acquired images. The spatial coherence length is defined as

$$d_c = \frac{\lambda_e}{2\alpha},$$ (14)

where λ_e is the electron wavelength and α is the RMS illumination convergence angle. Using the obtained illumination convergence angle, the spatial coherence length of the electron pulses generated with the RF gun is d_c = 5.6 nm, twice that of the current UED systems. This enables to detect sharp DPs and acquire good-contrast diffraction images with a single shot, as shown in **Figure 9**.

The temporal resolution can be characterized in the time-resolved electron diffraction studies. **Figure 10** presents an example of time-resolved experiment for observation of UED structural dynamics in a 35-nm thick single-crystalline silicon sample. The sample was pumped by a 400-nm laser beam with a pulse duration of 100 fs and density of 3.5 mJ/cm². The UED patterns were measured with a single shot using a 3-MeV electron pulse with the pulse duration of 100 fs containing 10⁵ electrons per pulse. The data elucidate the dynamics in the silicon crystals: the

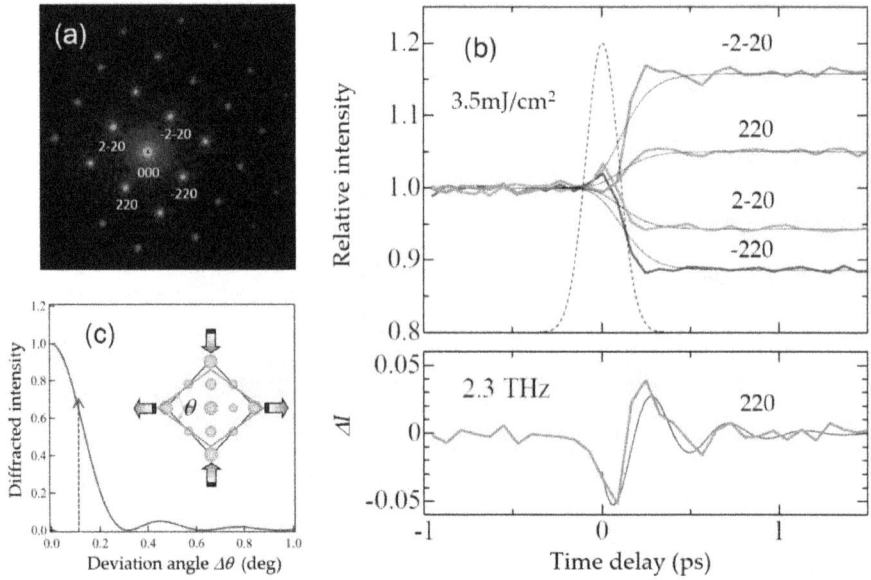

Figure 10.
UED structural dynamics. (a) DPs of a (100)-oriented single-crystalline silicon with a thickness of 35 nm measured by a 3-MeV femtosecond electron pulse with a single shot. The pusle duration of the electron pulse is ~ 100 fs containing 10^5 electrons. (b) Intensity changes for four Bragg spots after the laser illumination. The wavelength and density of the pump femtosecond laser are 400 nm and 3.5 mJ/cm2, respectively. (c) Simulation results of diffracted intensity changes due to lattice distortion within a angle of $\Delta\theta < 0.1°$ along the (111)-oriented direction [47].

diffracted intensities of the Bragg spots along the (−2–20) and (220) line increase, while the intensities of the Bragg spots along the (2–20) and (−220) line decrease after the laser pump. The structural recovery after a change is very slow. Tanimura and Naruse [47] suggested that these dynamics are caused mainly by the lattice distortion due to the laser excitation in silicon crystals, indicating that the time-resolved diffraction methodology in the relativistic UEM could be utilized to obtain an atomic perspective of lattice motion in crystalline materials.

The total temporal resolution of the UEM can be estimated mainly with the probe electron pulse duration σ_b, pump laser pulse width σ_b, and time jitter Δt_j between two pulses as

$$\Delta t = \sqrt{\sigma_b^2 + \sigma_l^2 + \Delta t_j^2}. \tag{15}$$

The time jitter is determined by the synchronization of the laser pulse to the accelerating RF phase; therefore, we can define $\Delta t_j = 61$ fs, described in Section 2.3. In our relativistic UEM, both pump and probe pulse durations are ~ 100 fs (RMS). The estimated total temporal resolution is approximately 150 fs (RMS), corresponding to the experimental results represented by the dashed line in **Figure 10(b)**.

4. Conclusions

In this chapter, we have introduced a relativistic UEM with femtosecond electron pulses, including the relativistic femtosecond electron pulse generation with an RF-acceleration-based photoemission gun, first prototype relativistic UEM instrument, and demonstrations of UEM image and UED measurements with the relativistic femtosecond electron pulses. The electron pulses generated by the RF gun exhibited excellent characteristics, including a low emittance of ≤ 0.1 mm-nrad, energy spread of 10^{-4}, and pulse duration of 100 fs with 10^6 electrons per pulse at an energy ≥ 3 MeV. These pulses have facilitated (1) the acquisition of high-quality high-resolution diffraction patterns with a single shot, (2) acquisition of acceptable UEM images with pulse integration measurements, and (3) time-resolved experiments with an excellent temporal resolution of 150 fs, suggesting that the relativistic UEM is very promising for studies on ultrafast phenomena in the femtosecond time region.

Ultrahigh-voltage electron microscopes operated in the relativistic-energy region above 1 MeV have attracted significant attention in various research fields, including physics, biology, and materials science. However, the existing system is too large and expensive to be purchased and operated even by large-scale facilities such as universities or national research centers, not to mention smaller research institutions and laboratories. The RF electron gun is capable of acceleration to ≥ 3 MeV in a length of only 15 cm. Further reductions in the emittance and energy spread, increase in the repetition rate, and suppression of some instabilities are required; however, the improvement in the RF gun and resolution refinement to the angstrom level will enable new electron microscopes based on this RF gun, sufficiently small and inexpensive for general research institutions and laboratories. Furthermore, by providing a femtosecond temporal resolution, such relativistic UEMs will constitute the next-generation electron microscopes, referred to as "dream machines," desired for a long time by material-structure researchers throughout the world.

Acknowledgements

The author acknowledges K. Kan, T. Kandoh, M. Gohdo, Y. Yoshida, H. Yasuda, and K. Tanimura of the Osaka University for their valuable suggestions and discussions. Additionally, the author thanks J. Urakawa, T. Takatomi, and N. Terunuma of the High Energy Accelerator Research Organization (KEK) for the design and fabrication of the high-quality RF gun.

This work was supported by a basic research (A) (No. 22246127, No. 26246026, and No. 17H01060) of the Grant-in-Aid for Scientific Research from the MEXT, Japan.

Author details

Jinfeng Yang
The Institute of Scientific and Industrial Research, Osaka University, Osaka, Japan

*Address all correspondence to: yang@sanken.osaka-u.ac.jp

IntechOpen

© 2018 The Author(s). Licensee IntechOpen. This chapter is distributed under the terms of the Creative Commons Attribution License (http://creativecommons.org/licenses/by/3.0), which permits unrestricted use, distribution, and reproduction in any medium, provided the original work is properly cited. [cc] BY

References

[1] Bostanjoglo O, Tornow RP, Tornow W. Nanosecond-exposure electron microscopy and diffraction. Journal of Physics E: Scientific Instruments. 1987;**20**:556-557. http://iopscience.iop.org/0022-3735/20/5/018

[2] Bostanjoglo O, Tornow RP, Tornow W. Nanosecond-exposure electron microscopy of laser-induced phase transformations. Ultramicroscopy. 1987;**21**:367-372. DOI: 10.1016/0304-3991(87)90034-9

[3] Bostanjoglo O, Heinricht F. A reflection electron microscope for imaging of fast phase transitions on surface. The Review of Scientific Instruments. 1990;**61**:1223-1229. DOI: 10.1063/1.1141952

[4] Bostanjoglo O, Elschner R, Mao Z, Nink T, Weingartner M. Nanosecond electron microscopes. Ultramicroscopy. 2000;**81**:141-147. DOI: 10.1016/S0304-3991(99)00180-1

[5] Kim JS, LaGrange T, Reed BW, Taheri ML, Armstrong MR, King WE, et al. Imaging of transient structures using nanosecond in situ TEM. Science. 2008;**321**:1472-1475. DOI: 10.1126/science.1161517

[6] LaGrange T, Campbell GH, Reed BW, Taheri M, Pesavento JB, Kim JS, et al. Nanosecond time-resolved investigations using the in situ of dynamic transmission electron microscope (DTEM). Ultramicroscopy. 2008;**108**:1441-1449. DOI: 10.1016/j.ultramic.2008.03.013

[7] Barwick B, Park HS, Kwon OH, Baskin JS, Zewail AH. 4D imaging of transient structures and morphologies in ultrafast electron microscopy. Science. 2008;**322**:1227-1231. DOI: 10.1126/science.1164000

[8] Zewail AH. Four-dimensional electron microscopy. Science. 2010;**328**:187-193. DOI: 10.1126/science.1166135

[9] Zewail AH, Thomas JM. 4D Electron Microscopy: Imaging in Space and Time. London: Imperial College Press; 2010. https://doi.org/10.1142/p641

[10] Piazza L, Masiel DJ, LaGrange T, Reed BW, Barwick B, Carbone F. Design and implementation of a fs-resolved transmission electron microscope based on thermionic gun technology. Chemical Physics. 2013;**423**:79-84. DOI: 10.1016/j.chemphys.2013.06.026

[11] Bücker K, Picher M, Crégut O, LaGrange T, Reed BW, Park ST, et al. Electron beam dynamics in an ultrafast transmission electron microscope with Wehnelt electrode. Ultramicroscopy. 2016;**171**:8-18. DOI: 10.1016/j.ultramic.2016.08.014

[12] Feist A, Bach N, Rubiano da Silva N, Danz T, Möller M, Priebe KE, et al. Ultrafast transmission electron microscopy using a laser-driven field emitter: Femtosecond resolution with a high coherence electron beam. Ultramicroscopy. 2017;**176**:63-73. DOI: 10.1016/j.ultramic.2016.12.005

[13] Kuwahara M, Nambo Y, Aoki K, Sameshima K, Jin X, Ujihara T, et al. The Boersch effect in a picosecond pulsed electron beam emitted from a semiconductor photocathode. Applied Physics Letters. 2016;**109**:013108. DOI: 10.1063/1.4955457

[14] Houdellier F, Caruso GM, Weber S, Kociak M, Arbouet A. Development of a high brightness ultrafast transmission electron microscope based on a laser-driven cold field emission source. Ultramicroscopy. 2018;**186**:128-138. DOI: 10.1016/j.ultramic.2017.12.015

[15] Manz S, Casandruc A, Zhang D, Zhong Y, Loch RA, Marx A, et al. Mapping atomic motions with ultrabright electrons: Towards fundamental limits in space-time resolution. Faraday Discussions. 2015;**177**:467-491. DOI: 10.1039/C4FD00204K

[16] Siwick BJ, Dwyer JR, Jordan RE, Dwayne Miller RJ. An atomic-level view of melting using femtosecond electron diffraction. Science. 2003;**302**:1382-1385. DOI: 10.1126/science.1090052

[17] Harb M, Ernstorfer R, Dartigalongue T, Hebeisen CT, Jordan RE, Dwayne Miller RJ. The Journal of Physical Chemistry. B. 2006;**110**:25308-25313. DOI: 10.1021/jp064649n

[18] Van Oudheusden T, de Jong EF, van der Geer SB, Op't Root WPEM, Luiten OJ. Electron source concept for single-shot sub-100 fs electron diffraction in the 100 keV range. Journal of Applied Physics. 2007;**102**:093501. DOI: 10.1063/1.2801027

[19] Sciaini G, Dwayne Miller RJ. Femtosecond electron diffraction: Heralding the era of atomically resolved dynamics. Reports on Progress in Physics. 2011;**74**:096101. DOI: 10.1088/0034-4885/74/9/096101

[20] Van Oudheusden T, Pasmans PLEM, Van der Geer SB, Loos MJD, Van der Wiel MJ, Luiten OJ. Compression of subrelativistic space-charge-dominated electron bunches for single-shot femtosecond electron diffraction. Physical Review Letters. 2010;**105**:264801. DOI: 10.1103/PhysRevLett.105.264801

[21] Hastings JB, Rudakov FM, Dowell DH, Schmerge JF, Cardoza JD, Castro JM, et al. Ultrafast time-resolved electron diffraction with megavolt electron beam. Applied Physics Letters. 2006;**89**:184109. DOI: 10.1063/1.2372697

[22] Li RK, Tang CX, Du YC, Huang WH, Du Q, Shi JR, et al. Experimental demonstration of high quality MeV ultrafast electron diffraction. The Review of Scientific Instruments. 2009;**80**:083303. DOI: 10.1063/1.3194047

[23] Musumeci P, Moody JT, Scoby CM. Relativistic electron diffraction at the UCLA Pegasus photoinjector laboratory. Ultramicroscopy. 2008;**108**:1450-1453. DOI: 10.1016/j.ultramic.2008.03.011

[24] Murooka Y, Naruse N, Sakakihara S, Ishimaru M, Yang J, Tanimura K. Transmission-electron diffraction by MeV electron pulses. Applied Physics Letters. 2011;**98**:251903. DOI: 10.1063/1.3602314

[25] Zhu P, Zhu Y, Hidaka Y, Wu L, Cao J, Berger H, et al. Femtosecond time-resolved MeV electron diffraction. New Journal of Physics. 2015;**17**:063004. DOI: 10.1088/1367-2630/17/6/063004

[26] Harb M, Peng W, Sciaini G, Hebeisen CT, Ernstorfer R, Eriksson MA, et al. Excitation of longitudinal and transverse coherent acoustic phonons in nanometer free-standing films of (001) Si. Physical Review B. 2009;**79**:094301. DOI: 10.1103/PhysRevB.79.094301

[27] Shen X, Li RK, Lundstrom U, Lane TJ, Reid AH, Weathersby SP, et al. Femtosecond mega-electron-volt electron microdiffraction. Ultramicroscopy. 2018;**184**:172-176. DOI: 10.1016/j.ultramic.2017.08.019

[28] Fu F, Liu S, Zhu P, Xiang D, Zhang J, Cao J. High quality single shot ultrafast MeV electron diffraction from a photocathode radio-frequency gun. The Review of Scientific Instruments. 2014;**85**:083701. DOI: 10.1063/1.4892135

[29] Giret Y, Naruse N, Daraszewicz SL, Murooka Y, Yang J, Duffy DM, et al.

Determination of transient atomic structure of laser-excited materials from time-resolved diffraction data. Applied Physics Letters. 2013;**103**:253107. DOI: 10.1063/1.4847695

[30] Daraszewicz SL, Giret Y, Naruse N, Murooka Y, Yang J, Duffy DM, et al. Structural dynamics of laser-irradiated gold nanofilms. Physical Review B. 2013;**88**:184101. DOI: 10.1103/PhysRevB.88.184101

[31] Yang J, Kan K, Naruse N, Yoshida Y, Tanimura K, Urakawa J. 100-femtosecond MeV electron source for ultrafast electron diffraction. Radiation Physics and Chemistry. 2009;**78**:1106-1111. DOI: 10.1016/j.radphyschem.2009.05.009

[32] Kan K, Yang J, Kondoh T, Yoshida Y. Development of femtosecond photocathode RF gun. Nuclear Instruments and Methods A. 2011;**659**:44-48. DOI: 10.1016/j.nima.2011.08.016

[33] Yang J, Yoshida Y, Shibata H. Femtosecond time-resolved electron microscopy. Electronics and Communications in Japan. 2015;**98**:50-57. DOI: 10.1002/ecj.11763

[34] Yang J. Ultrafast electron microscopy using relativistic-energy femtosecond electron pulses. Microscopy. 2015;**156-159**:50 (Japanese). http://microscopy.or.jp/jsm/wp-content/uploads/publication/kenbikyo/50_3/50_3e04jy.html

[35] Yang J, Yoshida Y, Yasuda H. Ultrafast electron microscopy with relativistic femtosecond electron pulses. Microscopy. 2018;in press. DOI: 10.1093/jmicro/dfy032

[36] Yang J, Kan K, Kondoh T, Yoshida Y, Tanimura K, Urakawa J. Femtosecond pulse radiolysis and femtosecond electron diffraction. Nuclear Instruments and Methods A. 2011;**637**:S24-S29. https://doi.org/10.1016/j.nima.2010.02.014

[37] Kurata H, Moriguchi S, Isoda S, Kobayashi T. Attainable resolution of energy-selecting image using high-voltage electron microscope. Journal of Electron Microscopy. 1996;**45**:79-84. DOI: 10.1093/oxfordjournals.jmicro.a023416

[38] Akre R, Dowell D, Emma P, Frisch J, Gilevich S, Hays G, et al. Commissioning the Linac coherent light source injector. Physical Review Accelerators and Beams. 2008;**11**:030703. DOI: 10.1103/PhysRevSTAB.11.030703

[39] Gulliford C, Bartnik A, Bazarov I, Cultrera L, Dobbins J, Dunham B, et al. Demonstration of low emittance in the Cornell energy recovery linac injector prototype. Physical Review Accelerators and Beams. 2013;**16**:073401. DOI: 10.1103/PhysRevSTAB.16.073401

[40] Terunuma N, Murata A, Fukuda M, Hirano K, Kamiya Y, Kii T, et al. Improvement of an S-band RF gun with a Cs_2Te photocathode for the KEK-ATF. Nuclear Instruments and Methods A. 2010;**613**:1-8. DOI: 10.1016/j.nima.2009.10.151

[41] Yang J, Sakai F, Yanagida T, Yorozu M, Okada Y, Takasago K, et al. Low-emittance electron-beam generation with laser pulse shaping in photocathode radio-frequency gun. Journal of Applied Physics. 2002;**92**:1608-1612. DOI: 10.1063/1.1487457

[42] Kim KJ. RF and space-charge effect in laser-driven RF electron guns. Nuclear Instruments and Methods A. 1989;**275**:201-208. DOI: 10.1016/0168-9002(89)90688-8

[43] Serafini L, Rosenzweig JB. Envelope analysis of intense relativistic quasilaminar beams in RF photoinjectors: A theory of emittance compensation. Physical Review E. 1997;**55**:7565-7590. DOI: 10.1103/PhysRevE.55.7565

[44] Travier C. An introduction to photo-injector design. Nuclear Instruments and Methods A. 1994;**340**:26-39. DOI: 10.1016/0168-9002(94)91278-5

[45] Yang J, Kan K, Kondoh T, Murooka Y, Naruse N, Yoshida Y, et al. An ultrashort-bunch electron RF gun. The Journal of the Vacuum Society of Japan. 2012;**42-49**(Japanese):55. DOI: 10.3131/jvsj2.55.42

[46] GPT code. General Particle Tracer. Available from: http://www.pulsar.nl/gpt

[47] Tanimura K. Femtosecond Time-Resolved Atomic Imaging. 2017. Available from: http://www.sanken.osaka-u.ac.jp/labs/aem/project.top.html201 [Accessed: 2018-08-05]

Chapter 2

Electron Microscopic Recording of Myosin Head Power and Recovery Strokes Using the Gas Environmental Chamber

Haruo Sugi, Tsuyosi Akimoto and Shigeru Chaen

Abstract

Despite extensive studies, the amplitude and the mode of the myosin head movement, coupled with ATP hydrolysis, still remain to be a matter for debate and speculation. To obtain direct information about the ATP-coupled myosin head movement, we prepared synthetic myosin filaments (myosin-myosin rod copolymer), in which myosin heads were position-marked with gold particles via antibodies to myosin heads and kept in hydrated, living state in the gas environmental chamber. ATP was applied to the specimen iontophoretically by passing the current to an ATP-containing microelectrode, and the ATP-induced myosin head movement was recorded with an imaging plate system under a magnification of 10,000×, with the following novel findings: (1) In the absence of ATP, myosin heads fluctuate around a definite neutral position. (2) In the absence of actin filaments, myosin heads move away from the bare region of myosin filaments (recovery stroke, average amplitude, 6 nm) on ATP application and return to the neutral position after exhaustion of ATP. (3) In the presence of actin filaments, the ATP-induced myosin head power stroke exhibits two different modes depending on mechanical conditions. (4) Myosin heads determine the direction of ATP-induced movement without being guided by actin filaments.

Keywords: gas environmental chamber, muscle contraction, myosin head power stroke, myosin head recovery stroke, iontophoretic ATP application

1. Introduction

In 1954, Huxley and Hanson [1] made a monumental discovery that muscle contraction is caused by relative sliding between actin and myosin filaments, which constitute hexagonal lattice structure within muscle fibers. Later, it has been found that (1) a muscle is a machine converting chemical energy derived from ATP hydrolysis into mechanical work and that (2) the sliding between actin and myosin filaments is produced by cyclic attachment-detachment between myosin heads extending from myosin filaments and corresponding myosin-binding sites in actin filaments [2]. **Figure 1** illustrates the most plausible sequence of the actin-myosin interaction in the muscle, producing force and motion in the muscle. In relaxed muscle, individual myosin heads (M) are in the state, M-ADP-Pi, and are detached

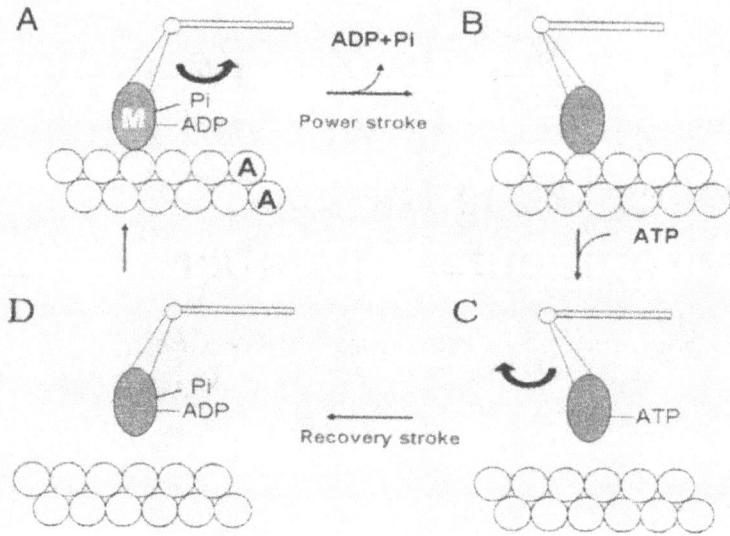

Figure 1.
Schematic diagram showing the most plausible sequence of attachment-detachment cycle between myosin head (M) and actin filament (A), coupled with ATP hydrolysis. From [7].

from actin filaments (A). On stimulation of the muscle, M-ADP-Pi attaches to A and performs a power stroke, associated with the release of Pi and ADP, to produce unitary sliding between actin and myosin filaments (from A to B). After completion of the power stroke, M remains attached to A until the next ATP comes to bind to it (B). On binding with ATP, M detaches from A, and performs a recovery stroke associated with ATP hydrolysis, to form M-ADP-Pi (from C to D). Then, M in the form of M-ADP-Pi again attaches to A to repeat power and recovery strokes.

Despite extensive studies to prove the power and recovery strokes associated with ATP hydrolysis, no definite results have been obtained due to the asynchronous nature of the individual myosin head movement in muscle fibers. In vitro motility assay experiments, in which fluorescently labeled actin filaments are made to slide on myosin molecules or their proteolytic fragments, and optical trap experiments, in which single myosin heads are made to interact with single actin filaments, are not effective in studying myosin head movements in the muscle, since these experiments differ too far from what actually takes place in the myofilament lattice structure constituting the muscle [3].

The most straightforward way to give information about the properties of myosin head power and recovery strokes is to visualize and record myosin head movements associated with ATP hydrolysis, using the gas environmental chamber (EC) attached to an electron microscope. The EC system enables us to insulate biological specimens from high vacuum of the electron microscope and to keep them in wet, living state by constantly circulating water vapor. We started to work with the group of the late Professor Akira Fukami of Nihon University, who developed the EC system together with the Japan Electron Optics Laboratory, Ltd. (JEOL Ltd.) in the late 1980s [4], and after a number of trials and errors, succeeded in obtaining novel results concerning the properties of myosin head power and recovery strokes.

Here, we describe novel properties of myosin heads in producing power and recovery strokes, which can never be obtained by any other experimental methods.

2. The EC system to record ATP-induced myosin head movement

As shown in **Figure 2**, the EC is a cylindrical compartment (inner diameter, 2.0 mm; depth, 0.8 mm) with upper and lower windows to pass electron beam. Each window is covered with a carbon sealing film (thickness, 15–20 nm) held on a copper grid with nine apertures (diameter, 0.1 mm). The carbon sealing film could bear pressure differences up to one atmosphere. The specimen, placed on the lower carbon film, is kept wet by constantly circulating water vapor (pressure, 60–80 torr; temperature 26–28°C). ATP is applied to the specimen by passing the current through an ATP-containing glass microelectrode. The EC was used in a 200 kV transmission electron microscope (JEM 2000EX; JEOL). The specimen images (magnification, 10,000×) are recorded with the imaging plate (IP) system (PIX system; JEOL). To avoid electron beam damage to the specimen, the total incident electron dose is limited below 5×10^{-4} C/cm^2. Further details have been described elsewhere [5, 6].

A myosin head consists of catalytic activation domain (CAD) and lever arm domain (LD), which are connected with a small converter domain (COD), and the LD is further connected to myosin subfragment-2 extending from the myosin filament backbone (**Figure 3**). Individual myosin heads were position-marked by colloidal gold particles (diameter, 20 nm) via three different antibodies: antibody 1 to the distal region of the CAD [7], antibody 2 to the proximal region of the CAD [7], and antibody 3 to two light chains of the LD [8]. Typical IP images of the bipolar synthetic filaments, with individual myosin heads positon-marked with gold particles, are presented in **Figure 4**. The specimen images could be recorded up to three times under a magnification of 10,000× without giving damage to the

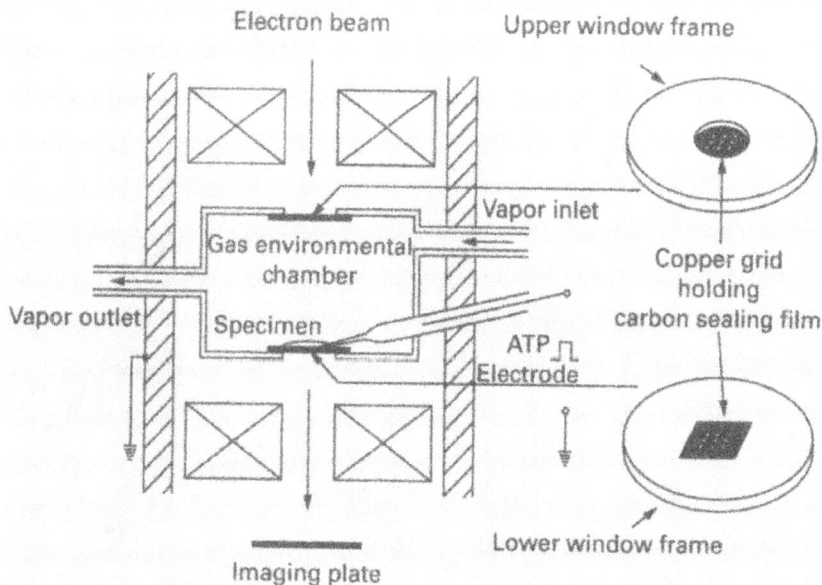

Figure 2.
Diagram showing the EC. The specimen (synthetic myosin filaments) is placed on the lower carbon sealing film and covered with thin layer of experimental solution. ATP is applied to the specimen by passing the current through the ATP-containing glass microelectrode. The specimen images are recorded with the imaging plate. From [6].

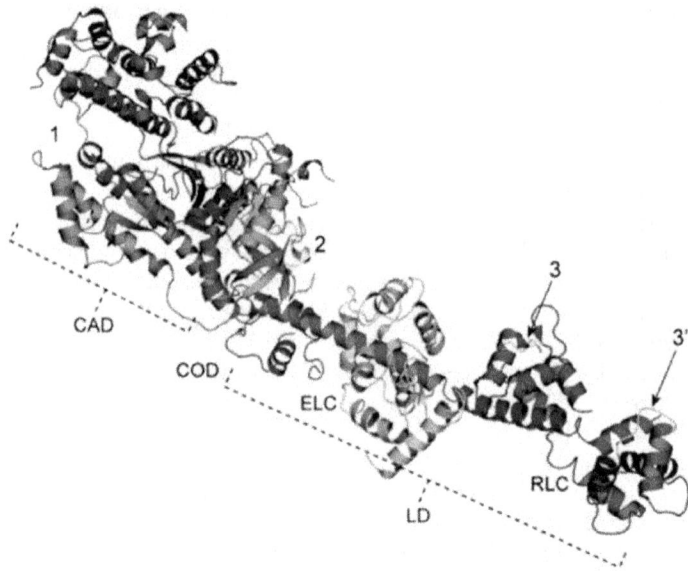

Figure 3.
Structure of a myosin head showing approximate regions of attachment of antibodies 1, 2, and 3, indicated by numbers 1, 2, and 3 and 3', respectively. The catalytic activation domain (CAD) consists of 25 K (green), 50 K (red), and 20 K (dark blue) fragments of myosin heavy chain, while the lever arm domain (LD) is composed of the rest of the 20 K fragment and essential (ELC, light blue) and regulatory (RLC, magenta) light chains. The CAD and LD are connected by the converter domain (COD). Location of peptides around Lys83 and that of two peptides, (Met58-Ala70) and (Leu106-Phe120), are colored yellow. From [9].

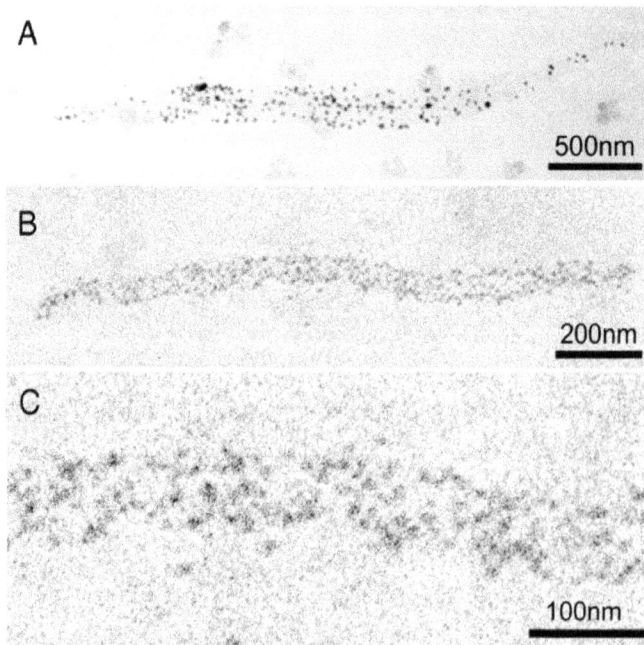

Figure 4.
(A, B) typical IP records of synthetic bipolar myosin filaments with a number of individual myosin heads, position-marked by gold particles. (C) Enlarged IP record showing part of the myosin filament in B. From [7].

specimen (exposure time, 0.1 s). In the above-recording condition, the pixel size was 2.5×2.5 nm, and the average number of electrons reaching each pixel during the exposure time was ~10. Reflecting the electron statistics, each gold particle image was composed of 20–50 pixels.

The center of mass position of each gold particle was determined as the coordinates (two significant figures) within a single pixel where the center of mass position was located. These coordinates representing the position of individual myosin heads were compared between two different IP records of the same myosin filaments. The distance D between the two centers of mass positions (with coordinates, X1 and X2 and Y1 and Y2, respectively) was calculated as D = $\sqrt{(X1—X2)^2 + (Y1—Y2)^2}$. Further details of the method of recording have been described elsewhere [5, 6].

3. Stable neutral position of myosin heads in the absence of ATP

Figure 5 shows the results of experiments, in which the image of the same myosin filament was recorded two times at an interval up to several minutes. It was found that the position of individual myosin heads remained almost unchanged

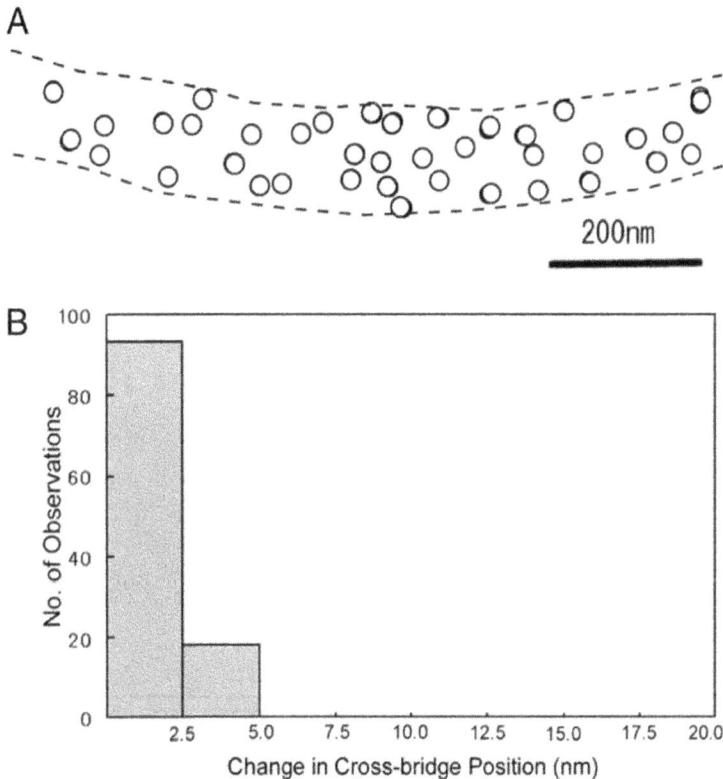

Figure 5.
Stable myosin head neutral position in the absence of ATP. (A) Difference between myosin head positions between two IP records of the same myosin filament on the common coordinates. Open and filled circles (diameter, 20 nm) were drawn around the center of mass position of each gold particle in the first and the second IP records, respectively. Note that filled circles are mostly covered by open circles because of nearly complete overlap between open and filled circles. In this and subsequent figures, broken lines indicate contour of myosin filament. (B) Histogram of distance between the two centers of mass positions of the same gold particle in the first and the second records. From [7].

with time, indicating that the position of each myosin head, time-averaged over 0.1 s (exposure time of the filament image to IP film), remains almost unchanged with time [6]. This result can be taken to indicate that each myosin head undergoes thermal motion around a definite neutral position, thus providing a favorable condition to detect ATP-induced myosin head movement.

4. ATP-induced recovery stroke in the absence of actin filament

On application of ATP to myosin filaments, individual myosin head was found to move in one direction parallel to the filament long axis with the average amplitude of ∼6 nm, as shown in **Figure 6** [7]. If the ATP-induced myosin head

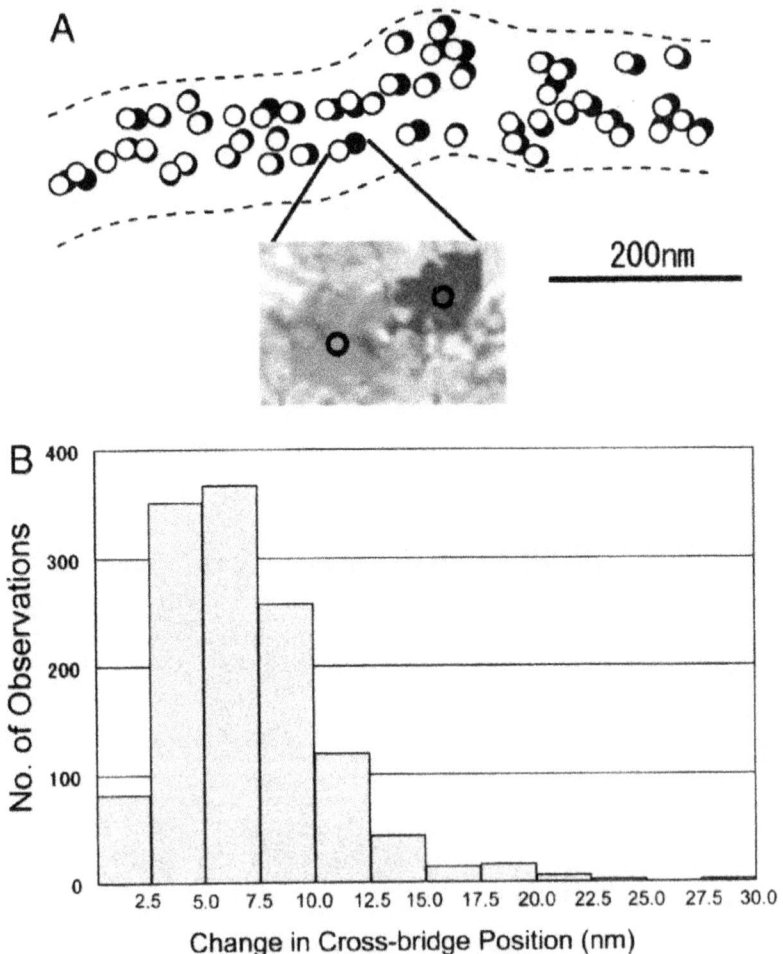

Figure 6.
ATP-induced myosin head movement in the absence of actin filaments. (A) Difference of myosin head position between the two IP records, one taken before, while the other taken after ATP application. Open and filled circles (diameter 20 nm) are drawn around the center of mass position of the same particles before and after ATP applications, respectively. (Inset) An example of superimposed IP records showing the change in position of the same gold particle, which are colored red (before ATP application) and blue (after ATP application). The center of mass position of each particle image is indicated by a small circle in each particle image. (B) Histogram of amplitude distribution of ATP-induced myosin head movement. From [7].

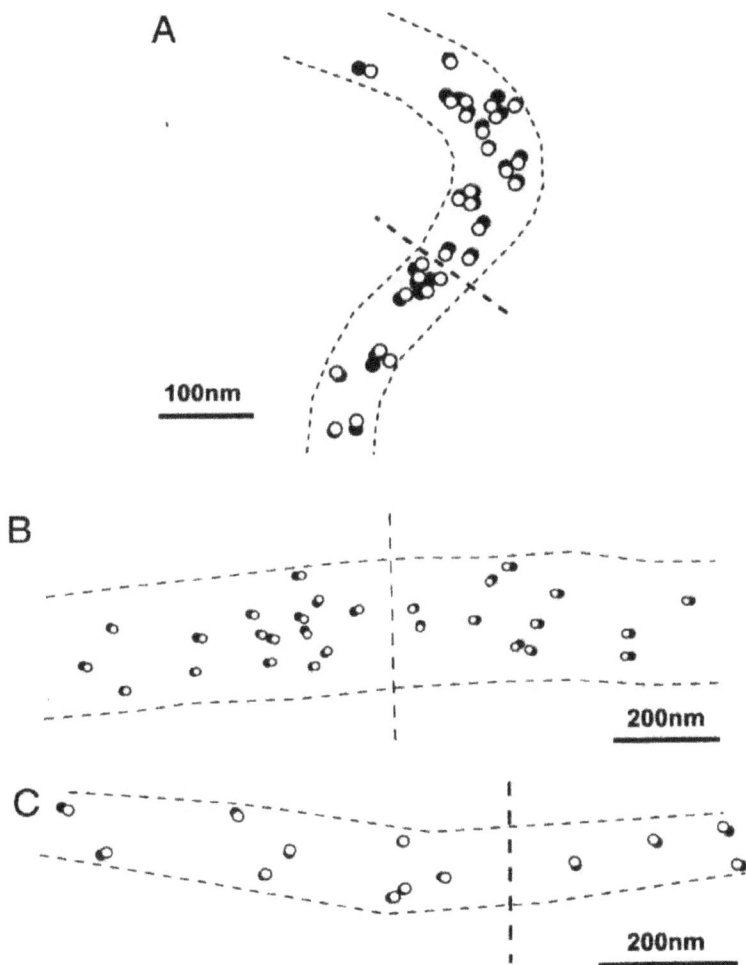

Figure 7.
(A–C) Examples of IP records showing the ATP-induced myosin head movement at both sides of myosin filament bare region, across which myosin head polarity is reversed. Open and filled circles (diameter, 20 nm) are drawn around the center of mass positions of the same particles in the IP records, taken before and after ATP applications, respectively. Note that myosin heads move away from the bare region. Approximate location of the bare region is indicated by broken lines across the center of myosin filament. From [7].

movement was recorded at both sides of the bare region, located at the center of myosin filament, individual myosin heads were found to move away from, but not toward, the bare region, across which the polarity of myosin heads is reversed (**Figure 7**) [7]. These findings indicate that, in the absence of actin filament, individual myosin heads perform a recovery stroke without being guided by actin filaments. In other words, each myosin head can sense the absence (or presence) of actin filaments to decide its direction of movement in response to ATP.

5. Return of myosin heads to the neutral position after exhaustion of ATP

In the absence of actin filaments, we took IP records of the same gold particles three times, i.e., (1) before ATP application, (2) during ATP application, and (3)

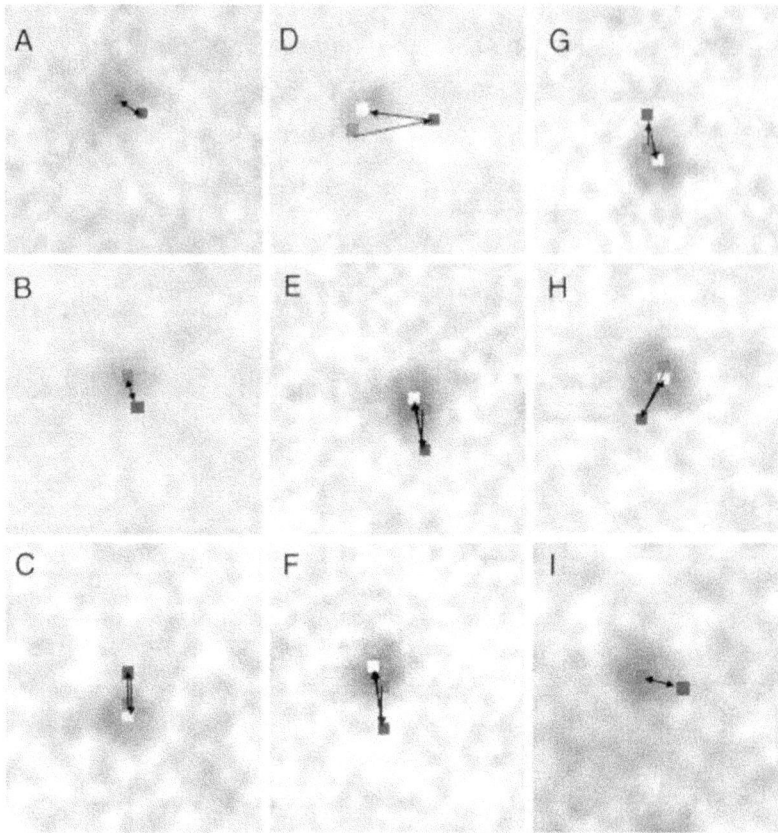

Figure 8.
(A–I) sequential changes in position of nine different pixels (each 2.5 × 2.5 nm), where the center of mass position of corresponding gold particles is located. In each frame, pixel positions before, during, and after complete exhaustion of ATP are indicated by red, blue, and yellow, respectively. Arrows indicate direction of myosin head movement. From [7].

after complete exhaustion of applied ATP (facilitated by adding hexokinase and D-glucose to experimental solution). **Figure 8** shows examples of sequential changes in position of nine different pixels (2.5 × 2.5 nm for each), where the center of mass positions of corresponding particles is located. It was found that myosin heads, which had performed recovery strokes, returned to or toward their initial neutral position after complete exhaustion of ATP. This can be taken to indicate that myosin heads can move in the direction similar to that of power stroke to return to their neutral position [7].

6. Amplitude of recovery stroke at three different regions within a myosin head

Our EC experiments can not only directly record myosin head movement, coupled with ATP hydrolysis but also record myosin head movement at different regions within a single myosin head by using three different antibodies to position-mark myosin heads [8]. As can be seen in the histograms of **Figure 9** A, B, and C, the amplitude of myosin head recovery stroke was the same in the distal region (6.14 ± 0.09 nm, mean ± SEM, n = 1692) and the proximal region (6.14 ± 0.22 nm,

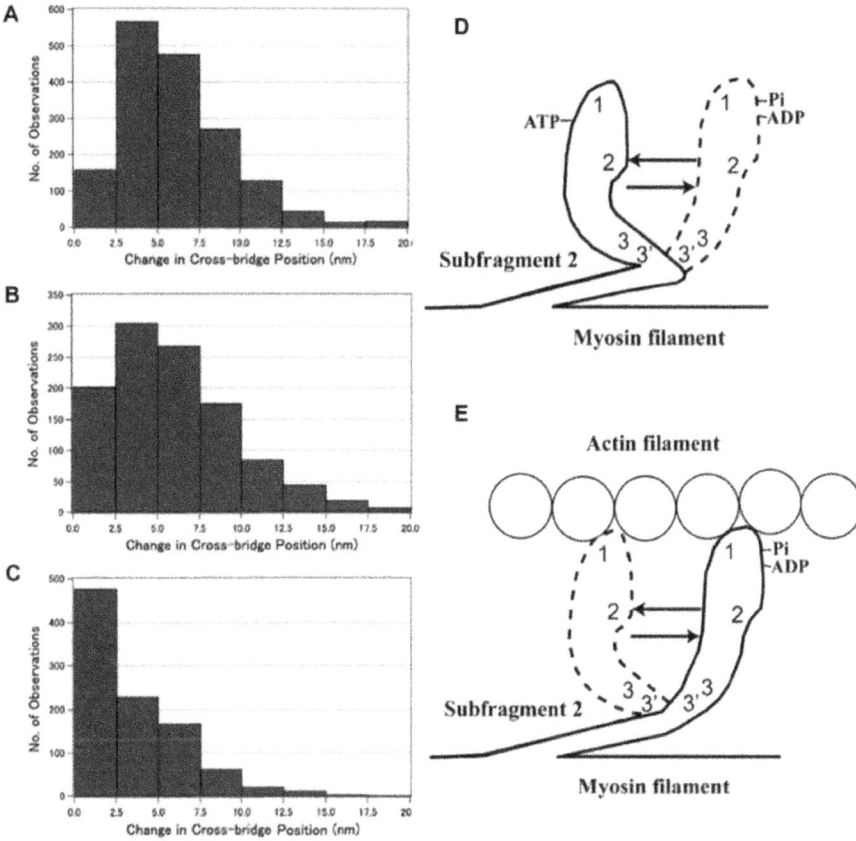

Figure 9.
(A–C) histograms of amplitude distribution of ATP-induced myosin head movement at the distal (A) and the proximal (B) regions of the myosin head CAD, at the two light chains in the proximal region of the myosin head LD. (D, E) Diagrams showing the ATP-induced configuration changes in the absence (D) and the presence (E) of actin filaments, based on the results shown in A–C. From [9].

n = 1112) of myosin head CAD, while it was much smaller at the LD region (3.55 ± 0.11 nm, n = 981) [9]. Based on these results, we can obtain diagrams shown in **Figure 9** D and E, in which myosin head CAD is perpendicular to actin and myosin filaments during the course of recovery stroke and subsequent return to their neutral position [9]. This view is consistent with our published results using the methods of quick-freezing and deep-etch replica that most myosin heads are perpendicular to actin and myosin filaments in relaxed, contracting, and rigor states [10].

7. Two different modes of myosin head power stroke in actin-myosin filament mixture

We also performed experiments, in which ATP-induced myosin head power stroke was recorded using the actin-myosin filament mixture, in which synthetic bipolar myosin filaments were surrounded by a number of actin filaments (**Figure 10**) [11]. Immediately, after mixing the actin and myosin filaments, the filament mixture, myosin heads form tight actin-myosin rigor linkages, some of which may be tension-bearing. Due to finite lifetimes of the rigor linkages [12],

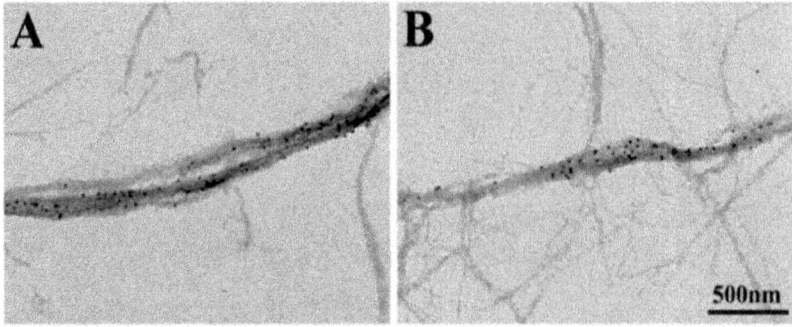

Figure 10.
Conventional electron micrographs of actin-myosin filament mixture. Myosin heads were position-marked with antibody 1 (left) and with antibody 2 (right), respectively. Note that bipolar myosin filaments are surrounded by actin filaments. From [11].

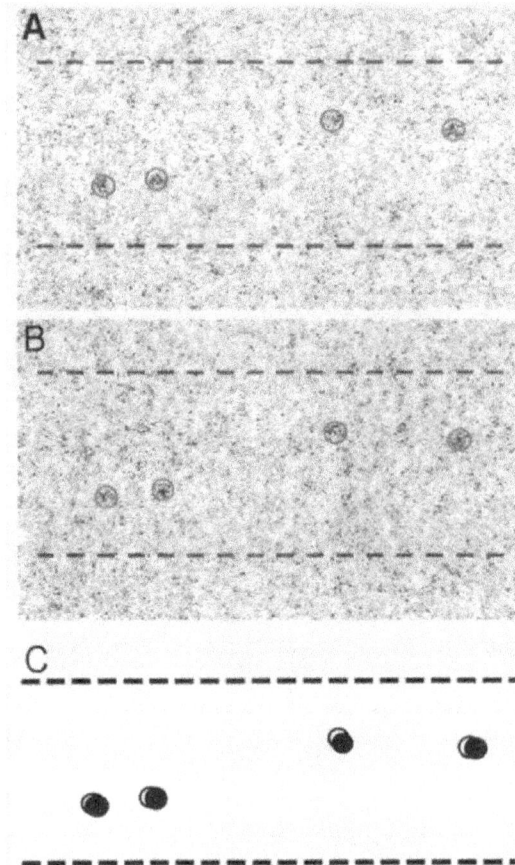

Figure 11.
(A, B) A pair of IP records of the same myosin filament mixture, taken before (A) and after (B) ATP application. Circles (diameter, 20 nm) are drawn around the center of mass position of individual gold particle images, consisting of a number of dark pixels. (C) Diagram showing ATP-induced changes in position of gold particles attached to individual myosin heads with antibody 1. Open and filled circles were drawn around the center of mass positions of the same particle before and after ATP applications, respectively. From [11].

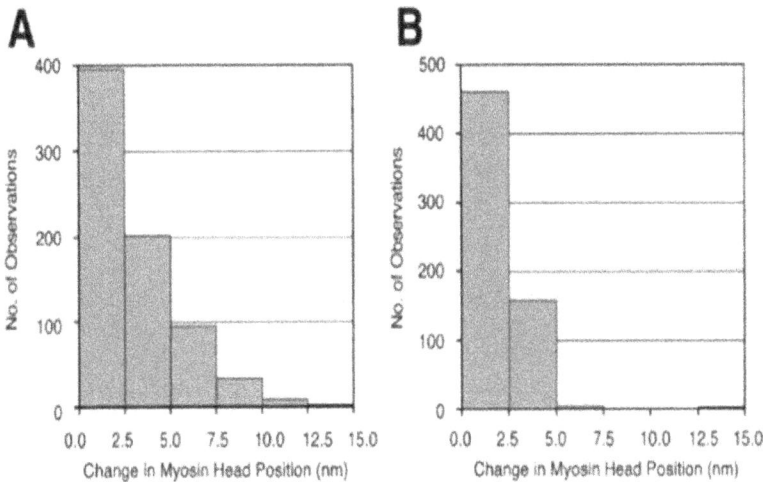

Figure 12.
Histograms of the amplitude distribution of ATP-induced myosin head power stroke at standard ionic strength at the distal (A) and the proximal (B) region of the myosin head CAD. From [11].

such rigor linkages may be broken with time, and myosin heads detached from actin first return to their neutral position and then again form rigor linkages with actin just opposite to them. As a result, all myosin heads are expected to form rigor linkages at their neutral position, where they exert no force, until the beginning of EC experiments at >20 min after the filament mixing.

Since the amount of ATP released from the ATP-containing electrode was limited, the ATP concentration around myosin filaments was estimated to be <10 µM. Reflecting this condition, only a small proportion of myosin heads could be activated by ATP, while the rest of them remained to form rigor linkages. Consequently, the ATP-activated myosin heads may not produce gross filament sliding but may only stretch adjacent elastic structures. **Figure 11** is a typical example of two IP records, taken before and after ATP applications. In accordance with this expectation, the amplitude of ATP-induced myosin head power stroke was small, being 3.3 ± 0.2 nm (mean ± SD, n = 732) at the distal region and 2.5 ± 0.1 nm (n = 613) at the proximal region of the myosin head CAD (**Figure 12**) [10]. If, however, the ionic strength of experimental solution was reduced from 125 to 20 mM, the amplitude of myosin head power stroke increased to 4.4 ± 0.1 nm (mean ± SD, n = 361) at the distal region and 4.3 ± 0.2 nm, n = 305) at the proximal region of the myosin head CAD (**Figure 13**) [10]. The increase in the amplitude of myosin head power stroke by reduction of ionic strength is consistent with our previous report that, at low ionic strength, the magnitude of Ca^{2+}-activated isometric tension in single-skinned fibers increases twofold, while the Mg-ATPase activity remains unchanged, indicating that the force generated by individual myosin heads increases twofold at low ionic strength [13].

Figure 14 shows diagrams showing two different modes of myosin head power stroke depending on experimental conditions. In the standard ionic strength, the amplitude of myosin head power stroke is larger at the distal region than at the proximal region, so that myosin head CAD is oblique to actin filament at the end of power stroke (A). At low ionic strength, the amplitude of myosin head power stroke is the same at both the distal and the proximal regions (B) [10]. This may be taken to imply that, under large loads, the myosin head CAD is oblique to actin filaments at the end of power stroke, while under moderate loads, the myosin head CAD is perpendicular to actin filaments.

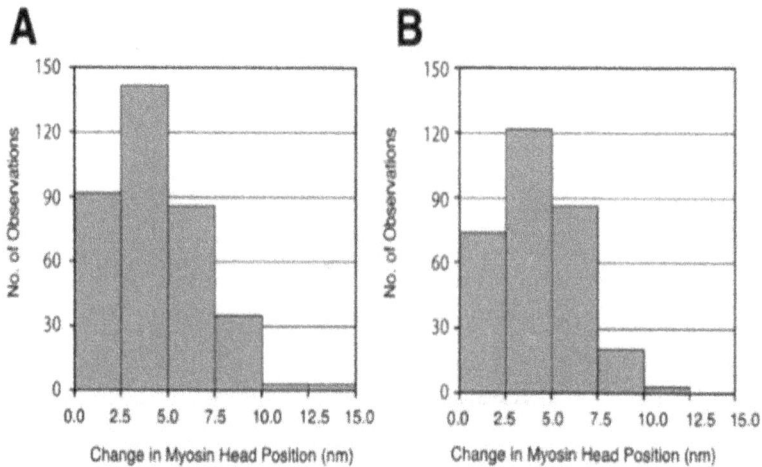

Figure 13.
Histograms of the amplitude distribution of ATP-induced myosin head power stroke at the distal (A) and the proximal (B) region of the myosin head CAD. From [11].

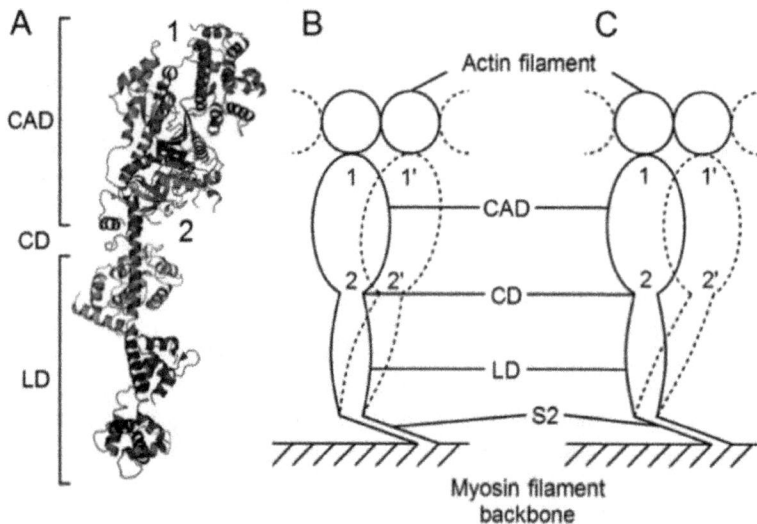

Figure 14.
Diagrams illustrating two different modes of myosin head power stroke, depending on experimental conditions. (A) Diagram of myosin head structure consisting of the CAD, COD, and LD. Approximate attachment regions of antibodies 1 and 2 are indicated by numbers 1 and 2, respectively. (B) The mode of myosin head power stroke at the standard ionic strength. Note that the amplitude of power stroke is larger at the distal region than at the proximal region of myosin head CAD. (C) The mode of myosin head power stroke at low ionic strength. Note that the amplitude of power stroke is the same at both the distal and the proximal regions of the myosin head CAD. From [11].

8. Conclusion

As has been described in this article, the experiments with the EC have the following advantages over any other methods to obtain information about myosin head power and recovery strokes, coupled with ATP hydrolysis: (1) it is possible to visualize and record ATP-induced movement in individual myosin heads within an electron microscopic field under a high magnification (10,000×); (2) using three different antibodies to myosin head, it is possible to record myosin head movement at three different regions within a myosin head, so that we can obtain information about changes in the mode of myosin head power stroke depending on experimental conditions. The EC experiments can be effectively used to obtain information not only about cardiac and smooth muscle myosins but also about other motile systems such as ciliary and flagellar movements. We hope that the EC experiments will be widely used to open new horizons in the research field of life sciences.

Acknowledgements

We wish to thank Presidents Kazuo Itoh, Kiichi Etoh, and Yoshiyasu Harada of Japan Electron Optics Laboratory (JEOL) for their generous support, without which our EC work would not have been achieved.

Author details

Haruo Sugi[1*], Tsuyosi Akimoto[1] and Shigeru Chaen[2]

1 Department of Physiology, School of Medicine, Teikyo University, Tokyo, Japan

2 Department of Integrated Sciences in Physics and Biology, College of Humanities and Science, Nihon University, Tokyo, Japan

*Address all correspondence to: sugi@kyf.biglobe.ne.jp

IntechOpen

© 2018 The Author(s). Licensee IntechOpen. This chapter is distributed under the terms of the Creative Commons Attribution License (http://creativecommons.org/licenses/by/3.0), which permits unrestricted use, distribution, and reproduction in any medium, provided the original work is properly cited. (cc) BY

References

[1] Huxley HE, Hanson J. Changes in the cross-striations of muscle during contraction and stretch and their structural interpretation. Nature. 1954;**173**:973-976

[2] Lymn RW, Taylor EW. Mechanism of adenosine triphosphate hydrolysis by actomyosin. Biochemistry. 1971;**16**:4617-4624

[3] Sugi H, Chaen S, Kobayashi T, Abe T, Kimura K, Saeki Y, et al. Definite differences between in vitro actin-myosin sliding and muscle contraction as revealed using antibodies to myosin head. PLoS One. 2014;**9**:e93272

[4] Fukami A, Adachi K. A new method of preparation of a self-perforated microplastic grid and its applications. Journal of Electron Microscopy. 1965;**14**:112-118

[5] Suda H, Ishikawa A, Fukam A. Evaluation of the critical electron dose on the contractile ability of hydrated muscle fibers in the film-sealed environmental cell. Journal of Electron Microscopy. 1992;**41**:223-229

[6] Sugi H, Akimoto T, Sutoh K, Chaen S, Oishi N, Suzuki S. Dynamic electron microscopy of ATP-induced myosin head movement in living muscle thick filaments. Proceedings of the National Academy of Sciences of the United States of America. 1997;**94**:4378-4382

[7] Sugi H, Minoda H, Inayoshi Y, Yumoto F, Miyakawa T, Miyauchi Y, et al. Direct demonstration of the cross-bridge recovery stroke in muscle thick filaments in aqueous solution by using the hydration chamber. Proceedings of the National Academy of Sciences of the United States of America. 2008;**45**:17396-17401

[8] Sutoh K, Tokunaga M, Wakabayashi T. Electron microscopic mapping of myosin head with site-directed antibodies. Journal of Molecular Biology. 1989;**206**:357-363

[9] Minoda H, Okabe T, Inayoshi Y, Miyakawa T, Miyauchi Y, Tanokura M, et al. Electron microscopic evidence for the myosin head lever arm mechanism in hydrated myosin filaments using the gas environmental chamber. Biochemical and Biophysical Research Communications. 2011;**405**:651-656

[10] Suzuki S, Oshimi Y, Sugi H. Freez-fracture studies on the cross-bridge angle distribution at various states and the thin filament stiffness in single skinned frog muscle fibers. Journal of Electron Microscopy. 1993;**42**:107-116

[11] Sugi H, Chaen S, Akimoto T, Minoda H, Minoda H, Miyakawa T, et al. Electron microscopic recording of myosin head power stroke in hydrated myosin filaments. Scientific Reports. 2015;**5**:17500

[12] Nishizaka T, Seo R, Tadakura H, Kinosita K, Ishiwata H. Characterization of single actomyosin rigor bonds: Load dependence of lifetime and mechanical properties. Biophysical Journal. 2000;**79**:962-974

[13] Sugi H, Kobayashi T, Chaen S, Ohnuki Y, Saeki Y, Sugiura S. Enhancement of force generated by individual myosin heads in skinned rabbit psoas muscle fibers at low ionic strength. PLoS One. 2013;**8**:e63658

Correlative Light-Electron Microscopy (CLEM) and 3D Volume Imaging of Serial Block-Face Scanning Electron Microscopy (SBF-SEM) of Langerhans Islets

Sei Saitoh

Abstract

Correlative light-electron microscopy (CLEM) is a developing technique for combined analysis of immunostaining for various biological molecules coupled with investigations of ultrastructural features of individual cells within a large field of view. This study first introduces a method of CLEM imaging of the same endocrine cells of compact and diffuse Langerhans islets from human pancreatic tissue specimens. The method utilises serial sections obtained from Epon-embedded specimens fixed with glutaraldehyde and osmium tetroxide. Next, serial block-face imaging using scanning electron microscopy (SBF-SEM) is advanced to enable rapid and efficient acquisition of three-dimensional (3D) ultrastructural information from Langerhans islets of mouse pancreas corresponding to the CLEM images. Samples for SBF-SEM observations were postfixed with osmium and stained en bloc and embedded in conductive resins with ketjenblack significantly reduced the charging of samples during SBF-SEM imaging.

Keywords: correlative light-electron microscopy (CLEM), serial block face SEM (SBF-SEM), compact and diffuse Langerhans islets, conductive resin

1. Introduction

Correlative light-electron microscopy (CLEM) is a developing technique for combined analysis of immunostaining for various biological molecules coupled with investigations of ultrastructural features of individual cells within a large field of view. Combined analysis of immunostaining for various biological molecules coupled with investigations of ultrastructural features of individual cells is a powerful approach for studies of cellular functions in normal and pathological conditions of human Langerhans islets.

Next, serial block-face imaging using scanning electron microscopy (SBF-SEM) is advanced to enable rapid and efficient acquisition of three-dimensional (3D)

ultrastructural information from Langerhans islets of mouse pancreas corresponding to the CLEM images of human. We confirmed the three-dimensional architecture of compact islets (tail) and diffuse islets (head) of the pancreas from normal
adult C57BL/6 J mice by SBF-SEM [1].

2. Langerhans islets have compact and diffuse type islets

A large number of endocrine cells, constituting 1–2% of the total volume of
the human pancreas in adults, are distributed in more than 1 million islets of
Langerhans, first described by Paul Langerhans in 1869. The pancreatic islets,
in turn, are distributed throughout the pancreas at variable densities in different lobules, although the density in the tail portion is usually slightly higher [2].
Pancreatic islet endocrine stem cells differentiated to mature islet endocrine cells
produce four major peptide hormones: insulin (β-cell granule), glucagon (α-cell
granule), somatostatin (δ-cell granule), and pancreatic polypeptide (PP-cell
granule), all of which show specific views by electron microscopy [3–5]. Human
and rodent pancreases are developed from two different embryological origins;
the dorsal origin develops the body and tail portions, whereas the ventral origin
derives the head portion of the pancreas. The majority of the pancreas is derived
from the dorsal anlage [3, 5–8]. The islets can be classified into two types [5].
"Compact" islets comprise the majority (90%); they are covered by nests of
connective tissues. Compact islets are composed of trabeculae of endocrine cells
interspersed with clear capsules between adjacent pancreatic exocrine acini
[9]. Other islets are "diffuse"; they have no nests separating them from adjacent
exocrine acini. However, the ultrastructure supporting "diffuse" pancreatic islets
including islet-encapsulating basement membranes, extracellular matrix, and
adjacent exocrine acini have not been elucidated in detail [5, 10–13]. Regenerating
islet-derived gene 1 alpha (REG1α) is secreted by the exocrine pancreas or β cells
and affects islet cell regeneration [14, 15], thereby regulating the development of
Langerhans islet architecture and diabetogenesis [10, 16, 17].

Conventional histochemical methods, such as aldehyde-fuchsin staining,
Hellerstrom-Hellman silver staining, and immunohistochemical labelling of
peptide hormones, are currently the major approaches used with light microscopy
to directly distinguish between types of endocrine cells [18, 19]. Although examination of multiple peptide contents in combination with investigation of the ultrastructural features of individual endocrine cells would provide a detailed analyses
of physiological and pathological alterations of pancreatic islets in genetically
and epigenetically divergent samples such as human tissues, correlative light and
electron microscopy observations combined with double immunostaining using
fluorescently labelled secondary antibodies in the same Epon-embedded sample is
an improved technique for correlative light-electron microscopy mapping which has
been described before [20–24].

3. Correlative light-electron microscopy (CLEM)

For the last quarter century, correlative microscopy, combining the power
and advantages of different imaging systems (light, electron, X-ray, NMR, etc.),
such as confocal laser scanning microscopy (CLSM), super-resolution microscopy (SFM), transmission electron microscopy (TEM) and scanning electron
microscopy (SEM), atomic force microscopy (AFM), magnetic resonance
imaging (MRI), superconducting quantum interference devices (SQUIDs), and

in vivo imaging (IVIS@) containing micro/nano CT (computed tomography), has become an important tool for biomedical research (**Figure 1**) [25–31]. In particular, the development of a series of hybrid approaches in technological advancements of microscopy techniques, labelling tools, and fixation or prepa-ration procedures allow correlating functional fluorescence microscopy data and ultrastructural information from a singular biological event. A key role in recent advancements of nanotechnology-based biomedical sciences is based on information obtained by light or electron microscopy. As correlative light electron microscopy (CLEM) approaches become increasingly accessible, long-standing questions of biology and clinical medicine regarding structure-function relation are being revisited [26, 31–36].

3.1 Sample preparation for CLEM

3.1.1 Fixation

For good observation of a biological sample in CLEM, fixation remains the ultrastructure of cells or tissue as close to the living material as possible, and sub-sequent dehydration and embedding. For light microscopy, the chemical fixation was originally designed to preserve the molecular structures of cells and tissues as well as the immunolocalization of components during the subsequent steps of preparation, such as alcohol dehydration and paraffin embedding [37–39]. On the other hand, double fixation with GA and $OsSO_4$ is suitable for EM observation of the ultrastructure of biomaterials. During the whole process of the fixation and embedding, tissue antigens undergo physicochemical modifications which results in masking of the mostly linear epitopes carried by the tissue components. For that reason, the fixative of immunoelectron microscopy (immuno-EM) is routinely limited to low concentration (0.05~0.5%) glutaraldehyde (GA) and formaldehyde (FA) before antigen-antibody reaction because osmium tetroxide (OsO4) markedly reduce antigen-antibody response [40, 41].

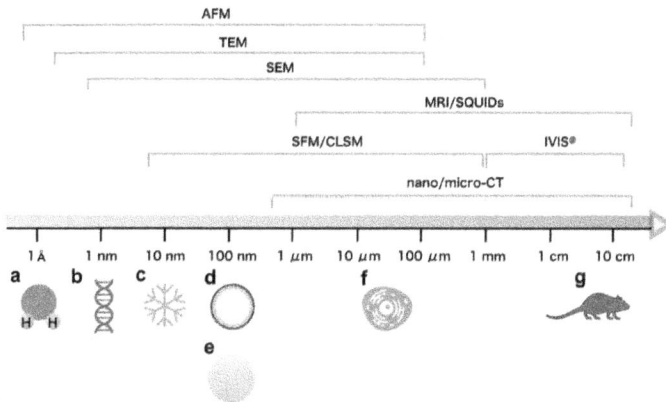

Figure 1.
Scale-based representative objects and corresponding microscopic tools for CLEM imaging: atomic force microscopy (AFM), transmission electron microscopy (TEM) and scanning electron microscopy (SEM), magnetic resonance imaging (MRI), superconducting quantum interference devices (SQUIDs), confocal laser scanning microscopy (CLSM), super-resolution microscopy (SFM), and in vivo imaging (IVIS@) containing micro/nano CT (computed tomography). (a) H_2O, water molecule (~2 Å), (b) DNA double helix (2–10 nm), (c) dendrimer (1–10 nm), (d) liposome (50–500 nm), (e) gold particle (50–200 nm), (f) cell (5–50 μm), and (g) a mouse (2–10 cm).

3.1.2 Antigen-antibody assay

The quality of correlative image matching critically depends on the ability to maintain the native organisation of cell or tissue during fixation and subsequent sample preparation. Basically, CLEM imaging based on immuno-EM is three different approaches on antigen-antibody assay: (I) serial sectioning, (II) pre-embedding, and (III) post-embedding because TEM imaging is necessary for section-based assay such as ultra-thick sectioning of the embedded samples [26, 35, 42, 43].

I. Serial sectioning

Serial sectioning for CLEM was reported by Baskin D.G et al. (1979), using immunocytochemistry with osmium-fixed tissues, and broadly used for bioscience and clinical medicine embedding in Epoxy resin [20–24, 45].

II. Pre-embedding

In the pre-embedding method, all of the immunostaining is done prior to embedding the tissue. For pre-embedding labelling, all of the immunostaining is done prior to embedding the tissue in resin for ultrathin sectioning on TEM or preparing the samples on SEM. Antigen-antibody reaction is limited by antibody penetration, as usually under 10 μm thickness.

III. Post-embedding

In the post-embedding method, the antigen-antibody reaction is performed on plastic or Tokuyasu cryosections after embedding [46, 47, 99].

3.1.3 Antigen masking

The chemistry of epitope masking itself has been poorly understood. The molecular mechanisms behind antigen masking was that the formaldehyde easily cross-linked amino residues of soluble and structure molecules [48], resulting in artificial changes of the molecular structures.

Antigen masking mechanisms are assumed mainly as follows: (I) molecular modifications of the antigen-carrying proteins upon fixation and embedding, (II) intramolecular modifications leading to antigen masking intrinsic to the protein, and (III) intermolecular effects on other proteins located in close contact with the antigen-bearing one.

3.1.4 Antigen retrieval

To obtain antigen-antibody reaction, some antigen retrieval techniques are frequently used such as enzyme treatment, quick freezing and freeze substitution, freeze-thaw technique, and heating by a microwave apparatus or a high-pressure oven [38, 39, 49–53]. Heat-induced antigen retrieval (HIAR) was developed as a method frequently used for LM and EM samples [49, 51, 53]. The antigen retrieval effect was assumed to be caused by breaks of the cross-linked molecules. Precise mechanisms of (HIAR) is that the extended polypeptides by heating are charged negatively or positively at basic or acidic pH and that an electrostatic repulsion force acts to prevent random entangling of polypeptides caused by hydrophobic attractive force and to expose antigenic determinants, during cooling process of HIAR solution. HIAR is a powerful tool to all types of immuno-EM for antigen-antibody assay [54–56].

3.2 CLEM using genetically labelled tag

Many biological functions depend critically upon fine details of cell and tissue molecular architecture that developed imaging technique revealing evolutionally. To overcome the limitation of immune assay and capture in vivo imaging and subsequent data acquisition, genetically labelled tag (GFP, mini-SOG, and APEX2) is applied broadly to intact living cells, tissues, and animal models (*Drosophila*, *Caenorhabditis elegans*, zebrafish, and rodents) for 3D CLEM imaging. GFP is converted to HRP-DAB reaction products by photoconversion or by immunolabelling with anti-GFP antibody. New generation of genetically labelled tags (mini-SOG and APEX2) are specialised for CLEM imaging [57–63, 97, 98].

I. Mini singlet oxygen generator (mini-SOG)

Mini-SOG is a small flavoprotein (106 amino acids) derived from *Arabidopsis* phototropin 2 capable of singlet oxygen production upon blue light irradiation to generate the polymerisation of diaminobenzidine into an osmiophilic reaction product resolvable by EM.

II. APEX2

APEX is an engineered peroxidase that functions as an electron microscopy tag and a promiscuous labelling enzyme for live-cell proteomics. APEX2 (enhanced ascorbate peroxidase) is an engineered peroxidase that catalyses DAB reaction to render target structures electron-dense.

3.3 CLEM using formalin-fixed paraffin-embedded sample (FFPE) for clinical medicine

Using haematoxylin and eosin staining or fluorescence immunostaining of paraffin sections of formalin-fixed paraffin-embedded sample (FFPE) for clinical medicine and simple low-vacuum scanning electron microscopy revealed a three-dimensional survey method for assessing cell/tissue architectures. The CLEM methods are applied widely to human biomaterial resources for clinical medicine [64, 65].

3.4 CLEM using glutaraldehyde and osmium tetroxide-fixed Epon-embedded samples for human Langerhans islets

This study is a developing method of correlative light and electron microscopy imaging of the tissue specimens. The method utilises serial sections obtained from Epon-embedded specimens fixed with glutaraldehyde and osmium tetroxide [1].

3.4.1 Tissue preparation

Small pieces of human and mouse pancreatic tissue were prefixed with 2.5% glutaraldehyde in 0.1 M phosphate buffer (PB; pH 7.4) for 1 h and postfixed with 1% osmium tetroxide in 0.1 M PB for 1 h. The specimens were routinely dehydrated by passing the tissue through a series of solutions with increasing ethanol concentrations and then embedded in Epon 812 epoxy resin. To examine the specimens, thick

Epon sections were first cut at 0.5-μm thickness and routinely stained with toluidine blue (TB). These sections were checked and trimmed to visualise Langerhans islets during the next step.

To identify compact and diffuse Langerhans islets, Epon sections of the human pancreas were prepared, routinely fixed with glutaraldehyde and OsO4, stained with TB, and observed with a light microscope (**Figure 3**). The compact islets were revealed as large collections of endocrine cells having round to oval shapes (**Figure 3A**, black arrowheads). Their nuclei have homogeneous chromatin patterns with nucleoli, and the cells have moderately light cytoplasm. In contrast, diffuse islets were composed of trabeculae of endocrine cells interspersed between adjacent acini (**Figure 2B**, red arrowheads).

3.4.2 Improved serial sectioning techniques

Then, ultrathin sections were cut at 70–80-nm thickness with a diamond knife on an ultramicrotome and mounted on Φ1 × 2 mm single-slit copper grids with a Formvar film covered by evaporated carbon (**Figure 2A**). Then, serial 0.5-μm thick sections were cut and put on MAS-coated glass slides (Matsunami Adhesive Slides, Matsunami Glass, Osaka, Japan) for subsequent immunohistochemical staining (**Figure 2B**).

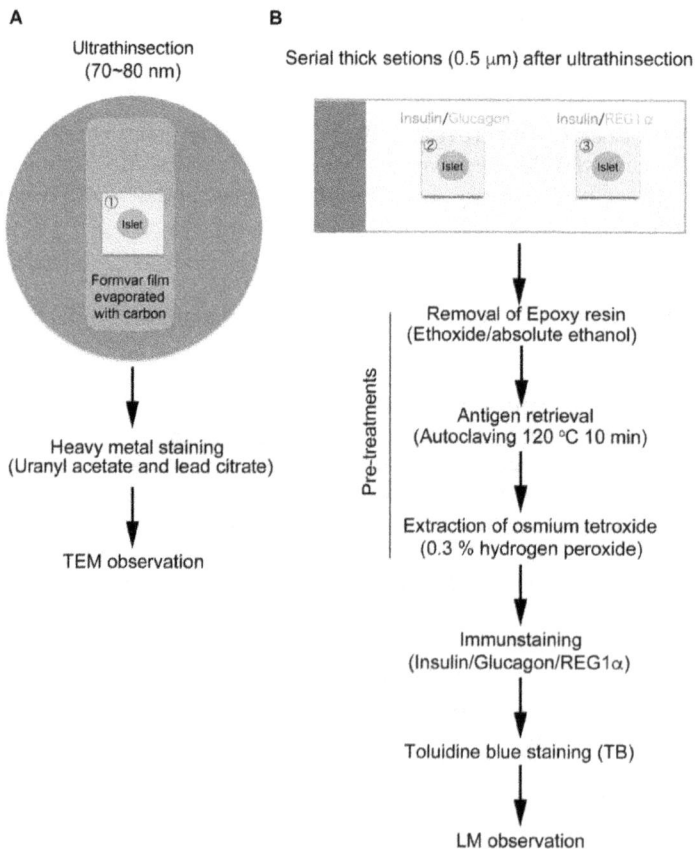

Figure 2.
Schematic representation of serial sectioning techniques for correlative light-electron microscopy mapping of human Langerhans islets. A: Schematic images illustrating an ultrathin section of a Langerhans islet mounted on a Φ1 × 2 mm single-slit copper grid with a Formvar film covered by evaporated carbon. B: Flow chart of pretreatments and immunohistochemical staining procedures applied to serial thick sections after ultrathin sectioning of Epon blocks.

The ultrathin sections of the human pancreas tissues on copper grids were double-stained with uranyl acetate and lead citrate and, finally, observed under a H-7500 transmission electron microscope (Hitachi, Tokyo, Japan) at an accelerating voltage of 80 kV. Electron microscopy images and montages of Langerhans islets were edited by Photoshop imaging software (Adobe Systems Incorporated, San Jose, CA, USA).

3.4.3 Immunohistochemistry in Epoxy thick sections

Immunohistochemistry in serial thick sections of Epon blocks. The 0.5-μm thick Epon sections on MAS-coated glass slides were placed on a heating plate and heated to 60–80°C for 15 min. During immunohistochemical staining of the peptide hormones (insulin and glucagon) and REG1α, we noted that two pretreatments after the removal of Epoxy resin, including antigen retrieval by autoclaving and extraction of osmium tetroxide with hydrogen peroxide (**Figure 3**).

3.4.3.1 Pretreatments

I. Removal of Epoxy resin

Epoxy resin was then removed from the sections by treatment with a mixture of ethoxide and absolute ethanol (1,2, v/v) for 30 min, washed in pure ethanol, and rehydrated in phosphate-buffered saline (PBS, pH 7.4). Prior to using the ethoxide/absolute ethanol mixture, saturated sodium ethoxide was aged for approximately 2 weeks until it turned dark brown [20].

II. Antigen retrieval by autoclaving

The specimens underwent two optional pretreatments. For antigen retrieval pretreatment, the specimens were autoclaved in 10 mM sodium citrate buffer (pH 6.0) at 120°C for 10 min and rinsed in PBS.

III. Extraction of osmium tetroxide

To extract osmium tetroxide, the specimens were placed in 0.3% hydrogen peroxide for 10 min and rinsed in PBS.

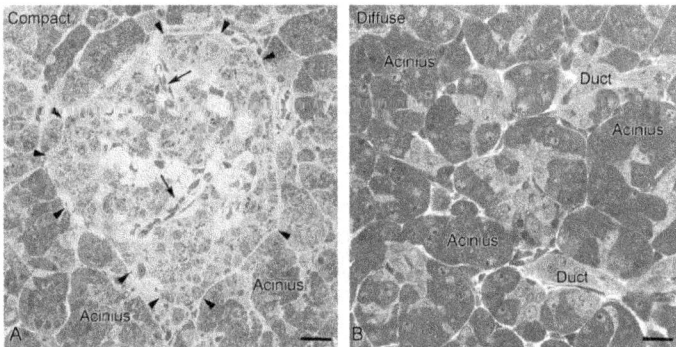

Figure 3.
Differences between compact and diffuse islets in human pancreas revealed by light microscopy observations of toluidine blue-stained thick sections routinely embedded in Epon. A: The compact islet appears round to oval and is composed of endocrine cells surrounded with a capsule of connective tissue (black arrowheads). Blood capillaries (black arrows) separate the islet into several lobules. Endocrine cells have moderately light cytoplasm. B: The diffuse islet is composed of a mass of endocrine cells interspersed between adjacent exocrine acinar-like cell clusters without a clear capsule (red arrowheads) in structures that appear like acinar ducts. Bars = 20 μm.

3.4.3.2 Immunoperoxidase-3,3'-diaminobenzidine (DAB) staining

For immunoperoxidase-3,3'-diaminobenzidine (DAB) staining, the specimens were treated with 3% fish gelatin (Sigma-Aldrich, St. Louis, MO, USA) in PBS for 10 min, followed by incubation with each primary antibody (insulin, glucagon, and REG1α) in PBS at 4°C overnight. Immunocontrols were performed using the same procedure with the exception that incubation with the primary antibody was omitted. The specimens were then incubated with a horseradish peroxidase (HRP)-conjugated secondary antibody in PBS for 1 h and visualised by exposure to metal-enhanced DAB (Pierce, Rockford, IL, USA) for 5 min. Finally, the specimens were incubated in 0.04% osmium tetroxide in 0.1 M PB for 30 sec to enhance contrast of the DAB reaction products.

The immunoreactivity of the HRP-DAB reaction was dramatically enhanced by autoclave treatment, and the effect was more prominent in insulin immunostaining (**Figure 4D** and **E**, insets; arrows) than in glucagon immunostaining (**Figure 4F** and **G**, insets; arrows). Immunocontrols not incubated with primary antibodies have reduced backgrounds caused by secondary antibody conjugated-HRP-DAB reaction products and osmification (**Figure 4B** and **C**).

3.4.3.3 Double-fluorescence immunostaining

For single- or double-fluorescence immunostaining experiments, de-osmified thick sections were treated with 3% fish gelatin in PBS for 10 min, followed by primary antibodies in PBS at 4°C overnight. Immunocontrols were performed using the same procedure with the exception that incubation with the primary antibody was omitted. As secondary fluorescently labelled antibodies, we used Alexa Fluor® 594 and Alexa Fluor 488-conjugated secondary antibodies. Immunostained sections were sealed with VECTASHIELD mounting medium (Vector Laboratories). Fluorescence signals for Alexa Fluor 488 or Alexa Fluor 594 were observed with a BX-61 fluorescence microscope (Olympus, Tokyo, Japan). After obtaining fluorescence images, some thick sections were additionally stained with TB for re-examination of morphology (**Figure 2B**).

Combined analyses using immunohistochemical localisation of hormones and ultrastructure of endocrine cells in animal pituitary glands have been previously reported (Baskin et al., 1979). In those studies, to achieve simultaneous examination, Epon thick sections were prepared from tissues fixed with glutaraldehyde and OsO_4 and peroxidase-DAB immunostaining was used for light microscopy, while electron microscopic observations were carried out in serial ultrathin sections [20]. Conventional double fixation was useful because lipid membranes, such as the plasma membrane of cells and vesicles, are well preserved. Epon-embedded sections are frequently used for ultrastructural analyses by electron microscopy because they exhibit well-preserved tissue morphology. However, the weak tissue antigenicity of Epon-embedded sections poses a problem for immunoassays. There have been attempts to improve immunolabelling of epoxy sections by etching hydrophobic Epoxy resin with different alkali solutions [66], retrieving antigenicity by sodium metaperiodate [21, 67], protease treatments [68], or heating (autoclaving or microwaving) thick or ultrathin sections for immunoelectron microscopy with various salt solutions [45, 56, 69–71]. In the present study, in addition to etching the hydrophobic Epon, we utilised autoclaving and pretreatment with hydrogen peroxide to enhance endocrine peptide immunoreactivity. It is believed that heating treatment retrieves immunoreactivity of masked antigens by exposing epitopes hidden because of cross-linking with aldehyde fixatives, whereas hydrogen peroxide treatment may increase immunoreactivity by reducing osmification of

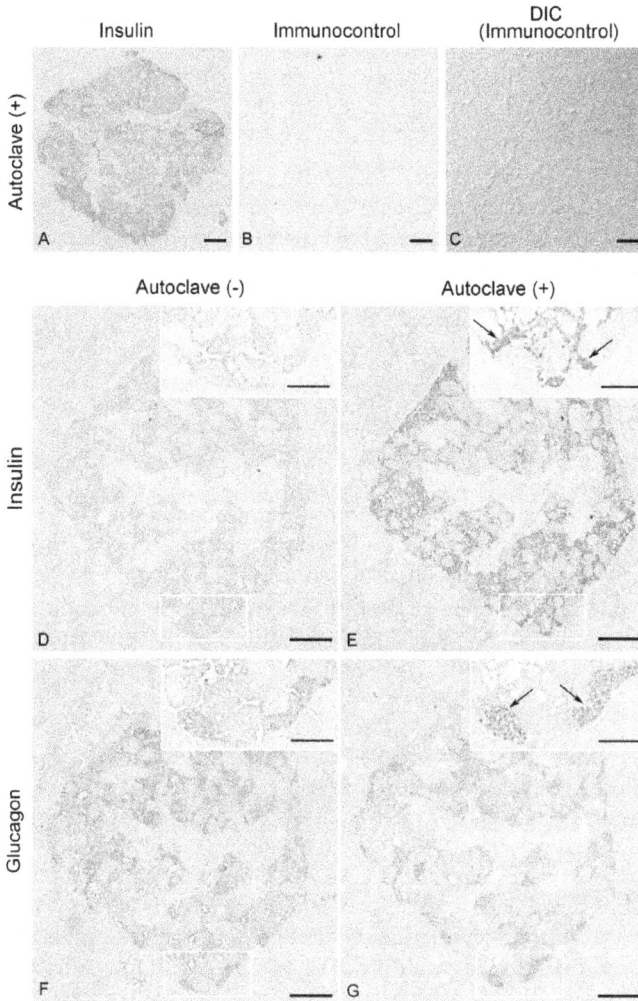

Figure 4.
Enhancement of immunostaining for insulin and glucagon in serial thick sections of a typical compact human pancreatic islet by autoclave treatment. A: Insulin-positive staining in a section that underwent autoclave treatment (AC). B: Immunocontrol incubated with secondary antibody without the primary antibody. C: Differential interference contrast microscope (DIC) image of the immunocontrol section. D: Weak insulin immunoreactivity in an untreated section. E: Stronger insulin immunoreactivity in a section that underwent AC—tiny granular patterns are more clearly detected throughout the cytoplasm of β cells (inset, arrows). F: Improved immunoreactivity for glucagon in the cytoplasm of cells in an untreated section. G: Clear granular pattern of immunoreactivity for glucagon in a section that underwent AC (inset, arrows). Bars = 20 μm (A–G); 10 μm (D–G, insets).

the target molecules [56, 69]. These pretreatments of Epon section are broadly applied to antibodies of immunohistochemistry not only for peptide hormones but also for several organelle markers: mitochondria, lysosome and peroxisome, or membrane proteins (aquaporin-1, aquaporin-2 and megalin) or soluble proteins— immunoglobulins (IgA and Ig kappa light chain), J chain, and albumin [45, 68, 71]. Overall, we demonstrated that a combination of double fixation, embedding in Epon, and immunohistochemistry with effective pretreatments was a very useful and robust approach for the simultaneous examination of cellular ultrastructure and antigen distribution in individual cells of the human pancreas. Points to be aware of regarding to pitfall or the limitations of this method: (I) carefully select for combination of first antibodies in double-fluorescence immunostaining, matching for the pretreatment conditions (heating or de-osmification) and (II) weaker DAPI staining for nucleus fluorescence staining after autoclaving [1].

4. CLEM revealed various types of secretory granules in individual endocrine cells of compact and diffuse Langerhans islets of the human pancreas

Correlative light and electron microscopy observations revealed various types of secretory granules in individual endocrine cells of compact and diffuse islets from specimens of the human pancreas (**Figures 5** and **7**).

4.1 Compact type of Langerhans islet

The use of our modified immunostaining protocol allowed clear observations of the ultrastructure of endocrine cells immunopositive for insulin, glucagon, and REG1α in serial thick and ultrathin sections in compact Langerhans islets of human pancreatic tissue (**Figure 5**). Immunostaining signals for insulin in β-cell granules and for glucagon in α-cell granules did not colocalize in the islet cells (**Figure 5A, D–F**), while patterns of immunostaining for insulin and REG1α overlapped in large parts of double-immunopositive endocrine cells (**Figure 5G–I**). Immunopositive staining for insulin and glucagon was positive for most islet areas whose ultrastructures were determined in serial ultrathin sections of the same islet (**Figure 5A, D–F, J, and K**). The compact islet illustrated in **Figure 4** had a higher ratio of cells positive for glucagon or insulin compared to cells from the compact islets shown in **Figure 3**. Areas of α or β cells on electron microscopy images were almost completely identical to the areas immunopositive for insulin and glucagon (**Figure 5A, D–F, J, and K**). These correlative observations revealed that islet cells with low and high immunoreactivity for glucagon included round-shaped granules with low and high electron density (**Figure 5P–R**; blue arrows), whereas insulin-positive cells with β-cell granules (**Figure 5O**; red arrows) and condensing vacuoles (**Figure 5O**, white arrows). Immunocontrols that were not incubated with primary antibodies have less background related to the Alexia 488- and Alexia 594-conjugated secondary antibodies and, in particular, because of secondary antibody cross-reactivity to native human immunoglobulins trapped in pancreatic tissues around the blood vessels and fixative autofluorescence.

Some endocrine cells of compact islets exhibited weak immunostaining for glucagon, and two types of round granules—with high and low electron densities—were observed in the same glucagon-positive cells identified in the corresponding areas of serial sections (**Figure 5**). During development and diabetogenesis, α cells may transdifferentiate into β cells for islet regeneration [21, 72]. Granules with a low electron density contain glucagon, glucagon-like peptide (GLP)-1, intervening peptide 2, GLP-2, and preproglucagon which is considered to be undergoing posttranslational processing because preproglucagon-containing granules are typically revealed as large α-cell granules with a lower electron intensity [3, 73–75]. In addition, some of the round granules with a low electron density may correspond to δ-cell granules, whose sizes are smaller than those of α-cell granules [76]. We consider that these δ-cell-like granules, which contain somatostatin, would potentially inhibit insulin and glucose secretion, express autocrine or paracrine to somatostatin receptors (SSTRs), and interact with the architecture of the islet [3, 77].

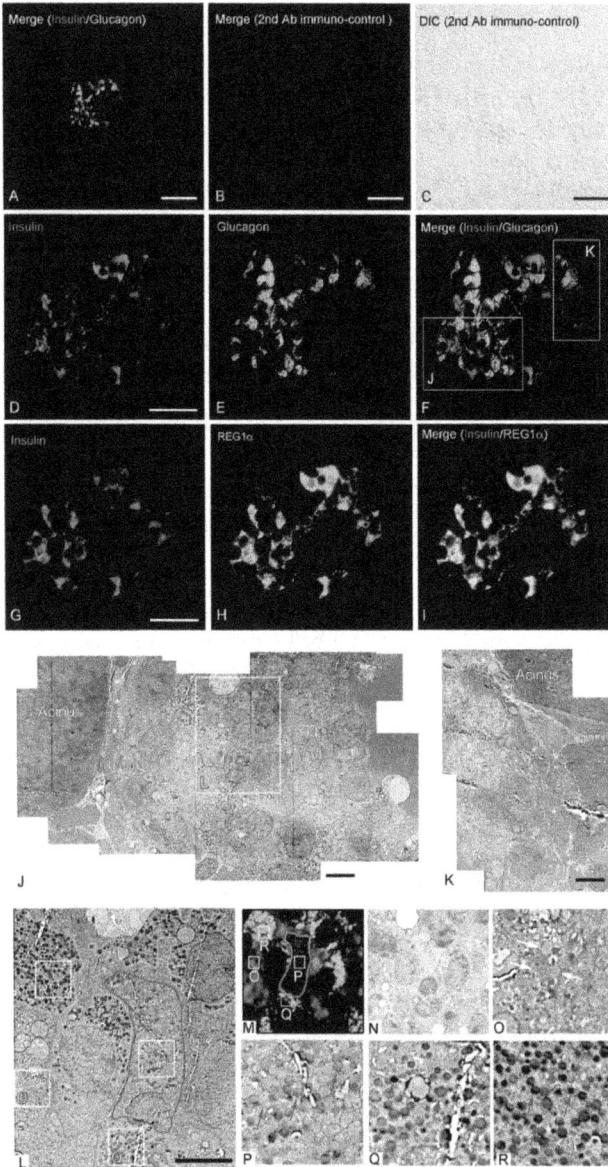

Figure 5.
Correlative light-electron microscopy mapping of a compact islet using serial Epon sections from a human pancreatic tissue specimen. Ultrastructural observations and double-immunofluorescence staining for insulin, glucagon, and REG1α were carried out. A: Merged image of insulin- and glucagon-positive staining. B: Merged image of immunocontrol (red and green) exposed to two secondary antibodies without primary antibodies. C: Differential interference contrast microscope (DIC) image of immunocontrol section shown in (B). Islet cells with granular immunopositive staining for insulin (A, D, F, G, I; red) are immunopositive for REG1α (H, I; green), but not for glucagon (A, E, F; green), whereas the immunostaining patterns of insulin and REG1α largely overlap (I; yellow). Electron microscopic montages (J, K) obtained from the same field of the serial section shown in (F). The magnified image (J, K) corresponds to the white boxes in (F). The compact islet has a capsule of connective tissue (J, K; yellow asterisks). Mapping images of β cell granules (J, K; pseudo-coloured red) and α-cell granules (J, K; pseudo-coloured green) manually coloured by Photoshop. Distribution patterns of insulin- (red) and glucagon-positive (green) cells (A, D–F) are clearly identified in the electron micrographs (J, K). Highly magnified electron microscopy images (O–R) correspond to the white boxes in (L) and double-immunopositive staining images (M). N: Toluidine blue staining for the same section (F) of insulin-positive cells (M; red) included β-cell granules (O; red arrows) and condensing vacuoles (O; white arrows), while glucagon-positive cells (M; green) included small numbers of round-shaped granules with low (P; blue arrows), with intermediate (Q; blue arrows) and high (R; blue arrows) electron densities. Bars = 20 μm (A–D, G); 1 μm (J–L).

4.2 Diffuse type of Langerhans islet

Immunoreactive staining for insulin and glucagon was also clearly observed in endocrine cells of diffuse islets, while exocrine cells of adjacent glandular acini were not immunostained (**Figures 6** and **7**). The immunoreactivities for REG1α and insulin showed different distributions in the same endocrine cells in serial thin sections of diffuse islets (**Figure 5B** and **C**). Fluorescently immunostained sections were subsequently stained with TB, and islet structures were found to be well preserved following double-fluorescence immunostaining (**Figure 7G** and **H**). Electron microscopy observations of serial ultrathin sections were performed to reveal structural details of diffuse islets in addition to identifying the hormones produced by the respective cells (**Figure 7**). It was found that β cells double-positive for REG1α and insulin were also the cells that exhibited zymogen-like condensing vacuoles (200–500 nm in size) and many organelles, such as mitochondria, Golgi apparatus, endoplasmic reticulum (ER), and lipofuscin granules (**Figure 7J** and **L**, white arrows) [78]. In addition, these endocrine cell granules in contact with exocrine acinar-like cell clusters have electron-dense cores and clear halos. However, insulin and REG1α double-positive endocrine cells consisted of several granular morphologies of human islet endocrine cells. We classified the granules of human islets into four types (α-, β-, δ-, and PP-cell granules) as described previously [3, 5, 74, 76] where (I) α-cell type (glucagon secretory), electron-dense without a clear halo occasionally presenting with a grey halo (**Figure 6K**, red arrowheads); (II) β-cell type (insulin secretory), granules of this type have electron-dense cores with a crystalline shape (**Figure 7K–N**, blue arrowheads); (III) δ-cell type (somatostatin secretary), larger and electron-opaque (**Figure 7K, L**, and **N**, cyan arrowheads); and (IV) PP-secretory cell type, spherical and smaller granules with a small halo (**Figure 7L**, green arrowheads). Interestingly, islet endocrine cells in contact with adjacent exocrine acinar-like cell clusters (ATLANTIS) contained zymogen-like granules (**Figure 7M**, yellow arrow), and cell-to-cell contacts were also detected (**Figure 7N**, white arrows).

Correlative light and electron microscopy analyses of serial thick and ultra-thin sections showed intracellular organelles and membrane interdigitations near cell-to-cell contact areas as well as typical α- or β-cell granules in individual insulin- and glucagon-positive endocrine cells located in both compact and diffuse islets (**Figures 5** and **7**). Human Langerhans islets contain polygonal endocrine cells that are demarcated by intercellular structures, such as tight junctions, gap junctions, and membrane structures, including interdigitations and invaginations [44]. In diffuse Langerhans islets, we found that some endocrine cells appear to

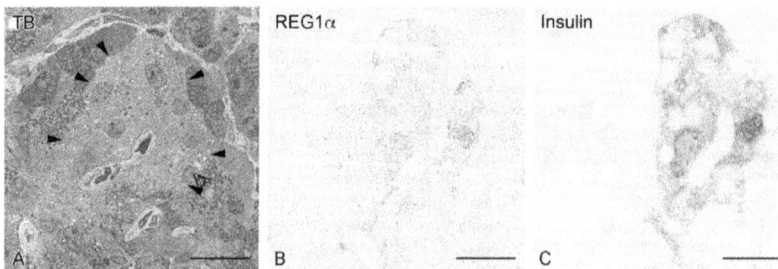

Figure 6.
Different distributions of HRP-DAB-immunoreaction products for REG1α and insulin in the same endocrine cells of diffuse type islet in serial thick sections of a diffuse human pancreatic islet. The diffuse islet is composed of a mass of endocrine cells interspersed between adjacent exocrine acinar-like cell clusters without a clear capsule (A; arrowheads). Islet endocrine cells show granular immunopositive staining for REG1α (B) or insulin (C), while acinar-like cells in contact with islet endocrine cells were not immunostained. TB: toluidine blue staining. Bars = 20 μm.

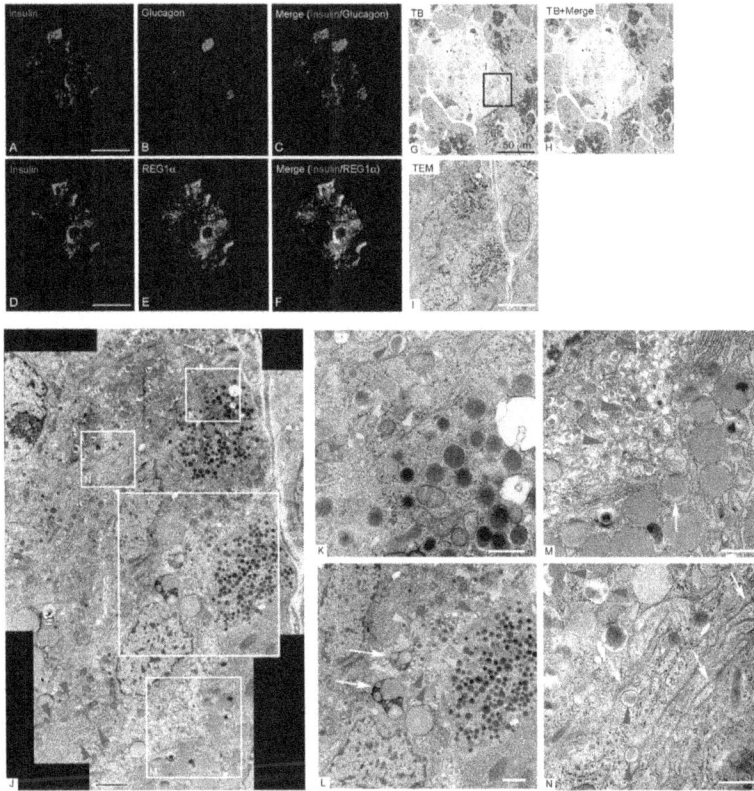

Figure 7.
Correlative light-electron microscopy mapping of a diffuse islet using serial Epon sections from a human pancreatic tissue specimen. Ultrastructural observation and double-immunofluorescence staining for insulin, glucagon, and REG1α were carried out. Granular immunoreactivities for insulin (A, C; red) and glucagon (B, C; green) are sparse and dense respectively, while the immunostaining patterns of insulin (D, F; red) and REG1α (E, F; green) largely overlap. G: A light microscopy image of the same thick section stained with toluidine blue (TB) after microscopy observation of fluorescence immunostaining (A–C). The diffuse islet is composed of a mass of endocrine cells in contact with adjacent exocrine acinar-like cell clusters without a clear capsule (red arrowheads). H: Superimposed image of (C) and (G). I: An electron microscopy image of the black-boxed area shown in (G) obtained from a serial ultrathin section demonstrating the ultrastructural features of insulin- and glucagon-positive endocrine cells in a diffuse pancreatic islet. The diffuse islet is composed of a mass of endocrine cells interspersed between adjacent exocrine acinar-like cell clusters without a clear capsule (red arrowheads). J: An electron microscopy montage showing a pancreatic islet corresponding to the rectangle of TB staining shown in (G). Endoplasmic reticulum and Golgi apparatus are indicated by blue and red arrowheads, respectively. K: A higher-magnification image of the respective boxed area in (J) illustrating an α cell containing α-cell granules (red arrowheads) in contact with a β cell containing β-cell granules (blue arrowheads) and condensing small vacuoles (cyan arrowheads). L: A higher-magnification image of the respective boxed area in (J) illustrating a β cell with β-cell granules (blue arrowheads) and spherical and smaller granules with small halo (green arrowheads) in contact with an α cell containing α-cell granules (red arrowheads). M: A higher-magnification image of the respective boxed area in (J) illustrating a β cell with β-cell granules (blue arrowheads) and a zymogen-like granule (yellow arrow) in cell-to-cell contact with an exocrine acinar-like cell containing zymogen-like granules. N: Interdigitation of cell membranes (white arrows) containing β-cell granules (blue arrowheads) and 200–500-nm condensing vacuoles (cyan arrowheads) between two β cells. Bars = 50 μm (A, D, G); 5 μm (I); 2 μm (J); 500 nm (K, N); 1 μm (L, M).

have direct cell-to-cell contacts with adjacent endocrine and exocrine acinar-like cells (**Figure 7**). These results indicate that electric or metabolic coupling exists not only between adjacent endocrine cells but also between endocrine cells and the surrounding pancreatic exocrine acinar-like cell clusters [10, 44]. We also identified insulin and REG1α double-positive cells that contained zymogen-like condensing granules in the endocrine islet cells, with several types of granules morphologically classified as β-, δ-, and PP-cell like granules (**Figure 6**).

5. Three-dimensional scanning electron microscopy (3D SEM) with volume imaging

Scanning electron microscopy (SEM) is a powerful technique for three dimensional CLEM imaging. SEM is traditionally used for imaging detecting the surface of cells, tissues, and the whole multicellular organisms by secondary electron beam. Recently, the critical advancement includes serial ultrastructural observation with scanning electron microscopy (SEM) using backscattered electron with specific tissue preparation methods to increase heavy metal deposition for efficient SEM imaging. In brief, three volume SEM methods: serial block-face electron microscopy (SBF-SEM), focused ion beam SEM (FIB-SEM), and array tomography using serial sectioning are illustrated in **Figure 8** [79–85, 96].

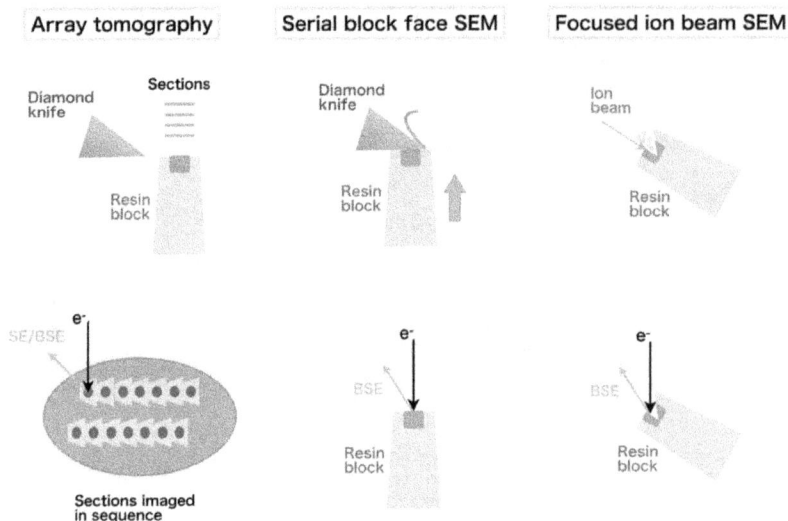

Figure 8.
A diagrammatic representation of three-dimensional scanning electron microscopy (3D SEM) techniques: array tomography (left), serial block face SEM (middle), and focused ion beam SEM (right). Array tomography is a volumetric microscopy method based on the ultrathin sections physically serial sectioned on an ultramicrotome and collected manually onto a substrate (e.g. a slide or coverslip) or onto a tape using an automated system, such as the ATUMtome. SBF-SEM consists of an ultramicrotome mounted automatically to cut the surface of resin block sample with diamond knife inside the vacuum chamber of a SEM. FIB-SEM is with an attached gallium ion column and the two beams. These electrons and ions (FIB) are focused on one coincident point of resin block sample.

6. Serial block-face scanning electron microscopy (SBF-SEM) revealed novel architecture of Langerhans islets

Serial block-face imaging using scanning electron microscopy (SBF-SEM) is advanced to enable rapid and efficient acquisition of three-dimensional (3D) ultrastructural information of large field of 3D volume imaging such as Langerhans islet over 100 μm size, providing a highly spatial resolution of the ultrastructure of diffuse islets from the head portion of mouse pancreas, which has a ventral origin [8, 86, 87].

6.1 Sample preparation using conductive Epon with carbon (ketjenblack)

Samples for SBF-SEM observations were postfixed with osmium and stained en bloc, as described previously. Briefly, mouse pancreatic tissues were prefixed with

2.5% glutaraldehyde in 0.1 M phosphate buffer (PB; pH 7.4) overnight, and tissues were washed with cacodylate buffer (pH 7.4). Notice that perfusion-fixation is easy to open the vessels and cell-to-cell contacts due to pressure artefacts. Tissues were then treated with 2% OsO4 (Nisshin EM, Tokyo, Japan) in 0.1 M cacodylate buffer containing 0.15% K4[Fe(CN)6] (Nacalai Tesque, Kyoto, Japan) for 1 h on ice and 0.1% thiocarbohydrazide (Sigma-Aldrich) for 20 min and 2% OsO4 for 30 min at room temperature. Thereafter, the tissues were treated with 2% uranyl acetate at 37°C for 3 h. Tissues were then treated with lead aspartate solution at 60°C for 30 min. The specimens were routinely dehydrated by passing the tissue through a series of solutions with increasing ethanol concentrations; infiltrated with acetone dehydrated with a molecular sieve, a 1:1 mixture of resin and acetone, and 100% resin; and then embedded in Epon 812 epoxy resin with carbon (ketjenblack) at 60°C for 3 days/overnight. Epon 812 epoxy resin with/without carbon (ketjenblack) enables for three-dimensional (3D) ultrastructural information of a large field of 3D volume imaging such as Langerhans islet over 100 μm size. Following trimming of islets from mouse pancreas, samples were imaged with a Sigma™VP (Carl Zeiss, Munich, Germany) equipped with 3View (Gatan Inc., Pleasanton, CA, USA). The serial images of SBF-SEM were handled with Fiji/ImageJ and segmented and reconstructed to 3D images using MIB (http://mib.helsinki.fi/) and Amira software.

Samples for SBF-SEM observations were postfixed with osmium and stained en bloc and embedded in conductive resins with ketjenblack significantly reduced the charging of samples during SBF-SEM imaging. Conductive resins were produced by adding the carbon black filler, ketjenblack, to resins commonly used for electron microscopic observations of biological specimens. Carbon black mostly localised around tissues and did not penetrate cells, whereas the conductive resins significantly reduced the charging of samples during SBF-SEM imaging. When serial images were acquired, embedding into the conductive resins improved the resolution of images by facilitating the successful cutting of samples in SBF-SEM [1, 88].

6.2 ATLANTIS in diffuse type of Langerhans islet

Endocrine cells from diffuse islets were also in contact with adjacent exocrine acinar-like cell clusters (ATLANTIS) without a clear capsule (**Figure 8D–I**). The morphologies of granules in endocrine cells of diffuse islets (**Figure 8F**) mainly consisted of three types: (I) granules with a spherical dense core and a small halo similar to PP-cell granules (**Figure 9F**, green arrowheads), (II) granules with a low electron-dense core without halo similar to δ-cell granules (**Figure 8F**, cyan arrowheads), and (III) granules with a spherical or crystal-shaped dense core with a clear halo similar to β-cell granules [87].

Serial SBF-SEM images revealed that zymogen-like granules are broadly distributed in these endocrine islet cells in contact with acinar-like cells. The zymogen-like granules have an isotropic distribution from ATLANTIS to the islet cells through direct contact with lamellar ERs in diffuse islets of the pancreatic head portion in adult normal mice derived from ventral origins (**Figure 9G–I**). Electron microscopic observations showed the same architecture in human diffuse islets in contact with acinar-like cells without a clear capsule. The typical diffuse endocrine islet in contact with acinar-like cells showed typical features of being rich in PP with a paucity of glucagon [3, 8]. Further, endocrine islet cells colocalised most of the PP-cell-like granules and zymogen-like granules directly through cell-to-cell contact sites with lamellar ERs, even if zymogen-like granules were excreted from the ATLANTIS. Using the correlative light and electron microscopy imaging described in the present study, additional basic and clinical studies for the precise

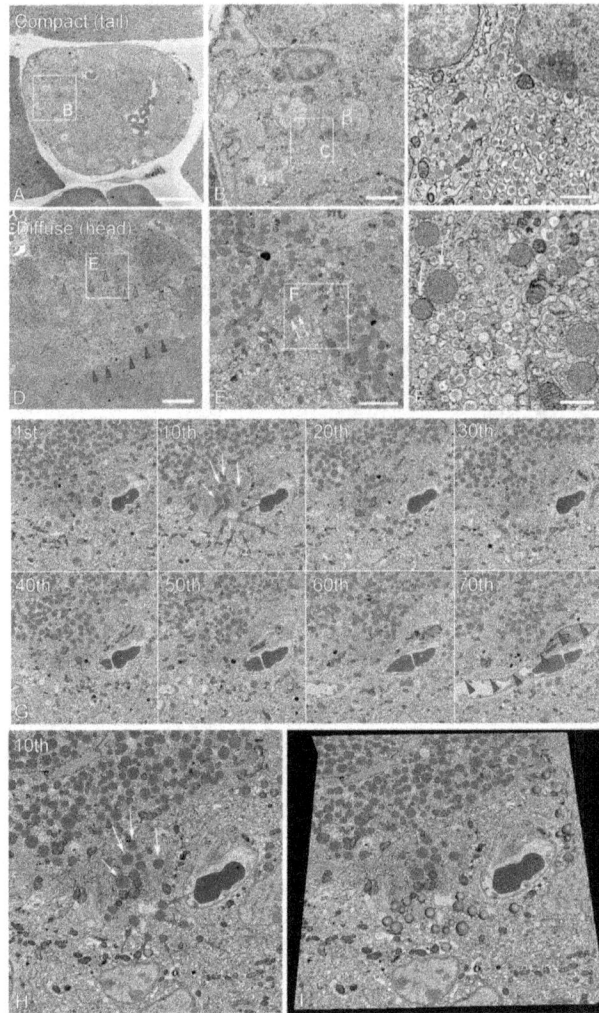

Figure 9.
Serial block-face scanning electron microscopy (SBF-SEM) observations of compact and diffuse islets from pancreatic tissues of a C57BL/6 J mouse. A: The compact islet is composed of a mass of endocrine cells with a clear capsule from the tail portion of the pancreas. B: α cells and β cells localised in the corner of an islet in contact with a capsule. C: α-cell granules (blue arrowheads) are electron-dense without a core, while β-cell granules (red arrowheads) show angular-shaped cores with a clear halo. D–E: The diffuse islet is composed of a mass of endocrine cells in contact with adjacent exocrine acinar-like cell clusters (ATLANTIS) (red arrowheads) without a clear capsule (blue arrowheads). F: The islet endocrine cells contain three type of granules: (1) spherical smaller granules with a small halo (green arrowheads), (2) slightly electron-opaque but spherical granules without a halo having a similar size to α-cell granules (cyan arrowheads), and (3) granules with zymogen-like granules (yellow arrows). The ATLANTIS containing β-cell granules (red arrowheads). G: Montage picture of the islet endocrine cells and ATLANTIS in the diffuse islet (D) showing every 10 serial images of a total of 70 images at 50-nm thickness each. The ATLANTIS containing zymogen-like granules (G10th, white arrows), isotopically localised to the islet cells (G10th, blue arrows), and in direct contact with endocrine cells with lamellar endoplasmic reticuli (G10th, red arrowheads). Islet cells are partially separated from the acinar-like cell clusters with connective tissues by blood vessels (G70th, blue arrowheads). H: Magnified image of (G10th). I: Three-dimensional reconstruction of serial images indicating the zymogen-like granules (shown in orange) inside the islet endocrine cells in contact with ATLANTIS.

identification of the observed granules in glucagon- and insulin-positive cells as well as the immunohistochemical detection of peptide components in human Langerhans islets for detailed clinical analyses of diabetes mellitus and chronic kidney or intestinal diseases closely related to metabolic disorders warrant further investigation [4, 5, 10, 16, 71, 78, 79, 89–93].

Interestingly, recently developed super-resolution microscopy (SFM) enables a detailed analysis distribution of biological molecules at an even higher resolution (e.g. a lateral resolution of 20–50 nm) by stochastic optical reconstruction microscopy when it is combined with new light microscopy technologies for nano-level analyses, an approach which may be applied to chemically fixed and Epon-embedded specimens [30, 33, 94, 95]. In combination with immunohistochemistry and in situ hybridization in Epon sections, the correlative microscopy observation method would be a more powerful approach capable of revealing human islet regeneration under genomic and transcriptome control such as SSTR expression in β cells from human islets [3, 70, 77].

Acknowledgements

This work was supported by the JSPS KAKENHI Grant Number 16 K08439 and by the EM facility in the National Institute for Physiological Sciences in Japan.

Competing interest statement

The authors declare that there are no conflicts of interest.

Author details

Sei Saitoh
Department of Anatomy II and Cell Biology, Fujita Health University School of Medicine, Toyoake, Japan

*Address all correspondence to: saitoh@fujita-hu.ac.jp

IntechOpen

© 2018 The Author(s). Licensee IntechOpen. This chapter is distributed under the terms of the Creative Commons Attribution License (http://creativecommons.org/licenses/by/3.0), which permits unrestricted use, distribution, and reproduction in any medium, provided the original work is properly cited. (cc) BY

References

[1] Saitoh S, Ohno N, Saitoh Y, Terada N, Shimo S, Aida K, et al. Improved serial sectioning techniques for correlative light-electron microscopy mapping of human Langerhans islets. Acta Histochemica et Cytochemica. 2018;**51**:9-20

[2] Wittingen J, Frey CF. Islet concentration in the head, body, tail and uncinate process of the pancreas. Annals of Surgery. 1974;**179**:412-414

[3] Brereton MF, Vergari E, Zhang Q, Clark A. Alpha-, delta- and PP-cells: Are they the architectural cornerstones of islet structure and co-ordination? The Journal of Histochemistry and Cytochemistry. 2015;**63**:575-591

[4] Briant L, Salehi A, Vergari E, Zhang Q, Rorsman P. Glucagon secretion from pancreatic α-cells. Upsala Journal of Medical Sciences. 2016;**121**:113-119

[5] Mills SE. Histology for Pathologists. 4th ed. Philadelphia: Wolters Kluwer Health/Lippincott Williams & Wilkins; 2012. pp. 777-816

[6] Kaung HC. Glucagon and pancreatic polypeptide immunoreactivities co-exist in a population of rat islet cells. Experientia. 1985;**41**:86-88

[7] Oikawa T, Ogawa K, Taniguchi K. Immunocytochemical studies on the pancreatic endocrine cells in the Japanese newt (*Cynopus pyrrhogaster*). Experimental Animals. 1992;**41**:505-514

[8] Suda K, Mizuguchi K, Hoshino A. Differences of the ventral and dorsal anlagen of pancreas after fusion. Acta Pathologica Japonica. 1981;**31**:583-599

[9] Grube D, Bohn R. The microanatomy of human islets of Langerhans, with special reference to somatostatin (D-) cells. Archivum Histologicum Japonicum. 1983;**46**:327-353

[10] Aida K, Saitoh S, Nishida Y, Yokota S, Ohno S, Mao X, et al. Distinct cell clusters touching islet cells induce islet cell replication in association with over-expression of Regenerating Gene (REG) protein in fulminant type 1 diabetes. PLoS One. 2014;**9**:e95110

[11] Bensley RR. Studies on the pancreas of the guinea pig. The American Journal of Anatomy. 1911;**12**:297-388

[12] Bonner-Weir S, Smith FE. Islets of Langerhans: Morphology and its implications. In: Kahn CR, Weir GC, editors. Joslin's Diabetes Mellitus. 13th ed. Philadelphia: Lea & Febiger; 1994. pp. 15-28

[13] Dewitt LM. Morphology and physiology of areas of Langerhans in some vertebrates. The Journal of Experimental Medicine. 1906;**8**:193-239

[14] Terazono K, Uchiyama Y, Ide M, Watanabe T, Yonekura H, Yamamoto H, et al. Expression of reg protein in rat regenerating islets and its co-localization with insulin in the beta cell secretory granules. Diabetologia. 1990;**33**:250-252

[15] Watanabe T, Yonekura H, Terazono K, Yamamoto H, Okamoto H. Complete nucleotide sequence of human reg gene and its expression in normal and tumoral tissues. The reg protein, pancreatic stone protein, and pancreatic thread protein are one and the same product of the gene. The Journal of Biological Chemistry. 1990;**265**:7432-7439

[16] Calderari S, Irminger JC, Giroix MH, Ehses JA, Gangnerau MN, Coulaud J, et al. Regenerating 1 and 3b gene expression in the pancreas of type 2 diabetic Goto-Kakizaki (GK) rats. PLoS One. 2014;**9**:e90045

[17] Xu W, Li W, Wang Y, Zha M, Yao H, Jones PM, et al. Regenerating

islet-derived protein 1 inhibits the activation of islet stellate cells isolated from diabetic mice. Oncotarget. 2015;**6**:37054-37065

[18] Brissova M, Fowler MJ, Nicholson WE, Chu A, Hirshberg B, Harlan DM, et al. Assessment of human pancreatic islet architecture and composition by laser scanning confocal microscopy. The Journal of Histochemistry and Cytochemistry. 2005;**53**:1087-1097

[19] Lumelsky N, Blondel O, Laeng P, Velasco I, Ravin R, McKay R. Differentiation of embryonic stem cells to insulin-secreting structures similar to pancreatic islets. Science. 2001;**292**:1389-1394

[20] Baskin DG, Erlandsen SL, Parsons JA. Immunocytochemistry with osmium-feed tissue. 1. Light microscopic localization of growth hormone and prolactin with the unlabeled antibody-enzyme method. The Journal of Histochemistry and Cytochemistry. 1979;**27**:867-872

[21] Bendayan M, Zollinger M. Ultrastructural localization of antigenic sites on osmium-fixed tissues applying the protein A-gold technique. The Journal of Histochemistry and Cytochemistry. 1983;**31**:101-109

[22] Nakai Y, Iwashita T. Correlative light and electron microscopy of the frog adrenal gland cells using adjacent Epon-embedded sections. Archivum Histologicum Japonicum. 1976;**39**:183-191

[23] Lacy PE. Electron microscopic identification of different cell types in the islets of Langerhans of the guinea pig, rat, rabbit and dog. The Anatomical Record. 1957;**128**:255-267

[24] Watari N, Tsukagoshi N, Honma Y. Correlative light and electron microscopy of the islets of Langerhans in some lower vertebrates.

Archivum Histologicum Japonicum. 1970;**31**:371-392

[25] Albers J, Pacilé S, Markus MA, Wiart M, Vande Velde G, Tromba G, et al. X-ray-based 3D virtual histology-adding the next dimension to histological analysis. Molecular Imaging and Biology. 2018;**20**:731-741. DOI: 10.1007/s11307-018-1246-3

[26] Begemann I, Galic M. Correlative light electron microscopy: Connecting synaptic structure and function. Frontiers in Synaptic Neuroscience. 2016;**8**:28. DOI: 10.3389/fnsyn.2016.00028

[27] Caplan J, Niethammer M, Taylor RM 2nd, Czymmek KJ. The power of correlative microscopy: Multi-modal, multi-scale, multi-dimensional. Current Opinion in Structural Biology. 2011;**21**:686-693

[28] Cortese K, Vicidomini G, Gagliani MC, Boccacci P, Diaspro A, Tacchetti C. High data output method for 3-D correlative light-electron microscopy using ultrathin cryosections. Methods in Molecular Biology. 2013;**950**:417-437

[29] Van Rijnsoever C, Oorschot V, Klumperman J. Correlative light-electron microscopy (CLEM) combining live-cell imaging and immunolabeling of ultrathin cryosections. Nature Methods. 2008;**5**:973-980

[30] Yin W, Mendenhall JM, Monita M, Gore AC. Three-dimensional properties of GnRH neuroterminals in the median eminence of young and old rats. The Journal of Comparative Neurology. 2009;**517**:284-295

[31] Wolff G, Hagen C, Grünewald K, Kaufmann R. Towards correlative super-resolution fluorescence and electron cryo-microscopy. Biology of the Cell. 2016;**108**:245-258

[32] Gibson KH, Vorkel D, Meissner J, Verbavatz JM. Fluorescing the electron:

Strategies in correlative experimental design. Methods in Cell Biology. 2014;**124**:23-54

[33] Huang B, Wang W, Bates M, Zhuang X. Three-dimensional super-resolution imaging by stochastic optical reconstruction microscopy. Science. 2008;**319**:810-813

[34] Hughes LC, Archer CW, ap Gwynn I. The ultrastructure of mouse articular cartilage: Collagen orientation and implications for tissue functionality. A polarised light and scanning electron microscope study and review. European Cells & Materials. 2005;**9**:68-84

[35] Jin SE, Bae JW, Hong S. Multiscale observation of biological interactions of nanocarriers: From nano to macro. Microscopy Research and Technique. 2010;**73**:813-823

[36] Robinson JM, Takizawa T, Pombo A, Cook PR. Correlative fluorescence and electron microscopy on ultrathin cryosections: Bridging the resolution gap. The Journal of Histochemistry and Cytochemistry. 2001;**49**:803-808

[37] Fox CH, Johnson FB, Whiting J, Roller PP. Formaldehyde fixation. The Journal of Histochemistry and Cytochemistry. 1985;**33**:845-853

[38] Saitoh S, Terada N, Ohno N, Ohno S. Distribution of immunoglobulin-producing cells in immunized mouse spleens revealed with "in vivo cryotechnique". Journal of Immunological Methods. 2008;**331**:114-126

[39] Saitoh S, Terada N, Ohno N, Saitoh Y, Soleimani M, Ohno S. Immunolocalization of phospho-Arg-directed protein kinase-substrate in hypoxic kidneys using in vivo cryotechnique. Medical Molecular Morphology. 2009;**42**:24-31

[40] Cortese K, Diaspro A, Tacchetti C. Advanced correlative light/electron microscopy: Current methods and new developments using Tokuyasu cryosections. The Journal of Histochemistry and Cytochemistry. 2009;**57**:1103-1112

[41] Oorschot VM, Sztal TE, Bryson-Richardson RJ, Ramm G. Immuno correlative light and electron microscopy on Tokuyasu cryosections. Methods in Cell Biology. 2014;**124**:241-258

[42] Nakane PK. States of the art of immunoelectron microscopy in Japan. Journal of Electron Microscopy. 1989;**38**(Suppl):S135-S141

[43] Oliver C. Pre-embedding labeling methods. Methods in Molecular Biology. 2010;**588**:381-386

[44] Orci L, Malaisse-Lagae F, Amherdt M, Ravazzola M, Weisswange A, Dobbs R, et al. Cell contacts in human islets of Langerhans. The Journal of Clinical Endocrinology and Metabolism. 1975;**41**:841-844

[45] Shimo S, Saitoh S, Saitoh Y, Ohno N, Ohno S. Morphological and immunohistochemical analyses of soluble proteins in mucous membranes of living mouse intestines by cryotechniques. Microscopy (Oxford). 2015;**64**:189-203

[46] Tokuyasu KT. A study of positive staining of ultrathin frozen sections. Journal of Ultrastructure Research. 1978;**63**:287-307

[47] Tokuyasu KT, Singer SJ. Improved procedure for immunoferritin labeling of ultrathin frozen sections. The Journal of Cell Biology. 1976;**71**:891-906

[48] Shi SR, Cote RJ, Taylor CR. Antigen retrieval techniques: Current perspectives. The Journal of Histochemistry and Cytochemistry. 2001;**49**:931-937

[49] Cattoretti G, Pileri S, Parravicini C, Becker MH, Poggi S, Bifulco C, et al. Antigen unmasking on formalin-fixed, paraffin-embedded tissue sections. The Journal of Pathology. 1993;**171**:83-98

[50] Ohno S, Terada N, Ohno N, Saitoh S, Saitoh Y, Fujii Y. Significance of 'in vivo cryotechnique' for morphofunctional analyses of living animal organs. Journal of Electron Microscopy. 2010;**59**:395-408

[51] O'Leary TJ, Mason JT. A molecular mechanism of formalin fixation and antigen retrieval. American Journal of Clinical Pathology. 2004;**122**:154; author reply 154-155

[52] Parajuli LK, Fukazawa Y, Watanabe M, Shigemoto R. Subcellular distribution of α1G subunit of T-type calcium channel in the mouse dorsal lateral geniculate nucleus. The Journal of Comparative Neurology. 2010;**518**:4362-4374

[53] Shi SR, Key ME, Kalra KL. Antigen retrieval in formalin-fixed, paraffin-embedded tissues: An enhancement method for immunohistochemical staining based on microwave oven heating of tissue sections. The Journal of Histochemistry and Cytochemistry. 1991;**39**:741-748

[54] Emoto K, Yamashita S, Okada Y. Mechanisms of heat-induced antigen retrieval: Does pH or ionic strength of the solution play a role for refolding antigens? The Journal of Histochemistry and Cytochemistry. 2005;**53**:1311-1321

[55] Yamanshita S. Heat-induced antigen retrieval: Mechanisms and application to histochemistry. Progress in Histochemistry and Cytochemistry. 2007;**41**:141-200

[56] Yamashita S, Okada Y. Heat-induced antigen retrieval in conventionally processed Epon-embedded specimens: Procedures and mechanisms. The

Journal of Histochemistry and Cytochemistry. 2014;**62**:584-597

[57] Martell JD, Deerinck TJ, Sancak Y, Poulos TL, Mootha VK, Sosinsky GE, et al. Engineered ascorbate peroxidase as a genetically encoded reporter for electron microscopy. Nature Biotechnology. 2012;**30**:1143-1148

[58] Lam SS, Martell JD, Kamer KJ, Deerinck TJ, Ellisman MH, Mootha VK, et al. Directed evolution of APEX2 for electron microscopy and proximity labeling. Nature Methods. 2015;**12**:51-54

[59] Ou HD. Visualizing viral protein structures in cells using genetic probes for correlated light and electron microscopy. Methods. 2015;**90**:39-48

[60] Shu X, Lev-Ram V, Deerinck TJ, Qi Y, Ramko EB, Davidson MW, et al. A genetically encoded tag for correlated light and electron microscopy of intact cells, tissues, and organisms. PLoS Biology. 2011;**9**:e1001041

[61] Sonomura T, Furuta T, Nakatani I, Yamamoto Y, Unzai T, Matsuda W, et al. Correlative analysis of immunoreactivity in confocal laser-scanning microscopy and scanning electron microscopy with focused ion beam milling. Frontiers in Neural Circuits. 2013;**7**:1-7

[62] Tsang TK, Bushong EA, Boassa D, Hu J, Romoli B, Phan S, et al. High-quality ultrastructural preservation using cryofixation for 3D electron microscopy of genetically labeled tissues. eLife. 2018. DOI: 10.7554/eLife.35524

[63] Qi YB, Garren EJ, Shu X, Tsien RY, Jin Y. Photo-inducible cell ablation in *Caenorhabditis elegans* using the genetically encoded singlet oxygen generating protein miniSOG. Proceedings of the National Academy of Sciences of the United States of America. 2012;**109**:7499-7504

[64] Onouchi T, Shiogama K, Mizutani Y, Takaki T, Tsutsumi Y. Visualization of neutrophil extracellular traps and fibrin meshwork in human fibrinopurulent inflammatory lesions: III. Correlative light and electron microscopic study. Acta Histochemica et Cytochemica. 2016;**49**:141-147

[65] Sawaguchi A, Kamimura T, Yamashita A, Takahashi N, Ichikawa K, Aoyama F, et al. Informative three-dimensional survey of cell/tissue architectures in thick paraffin sections by simple low-vacuum scanning electron microscopy. Scientific Reports. 2018;**8**:7479. DOI: 10.1038/s41598-018-25840-8

[66] Maxwell MH. Two rapid and simple methods used for the removal of resins from 1.0 micron thick epoxy sections. Journal of Microscopy. 1978;**112**:253-255

[67] Ozawa H, Picart R, Barret A, Tougard C. Heterogeneity in the pattern of distribution of the specific hormonal product and secretogranins within the secretory granules of rat prolactin cells. The Journal of Histochemistry and Cytochemistry. 1994;**42**:1097-1107

[68] Haraguchi CM, Yokota S. Immunofluorescence technique for 100-nm-thick semithin sections of Epon-embedded tissues. Histochemistry and Cell Biology. 2002;**117**:81-85

[69] D'Amico F, Skarmoutsou E, Stivala F. State of the art in antigen retrieval for immunohistochemistry. Journal of Immunological Methods. 2009;**341**:1-18

[70] Matsui T, Onouchi T, Shiogama K, Mizutani Y, Inada K, Fuxun Y, et al. Coated glass slides TACAS are applicable to heat-assisted immunostaining and in situ hybridization at the electron microscopy level. Acta Histochemica et Cytochemica. 2015;**48**:153-157

[71] Zhai XY, Kristoffersen IB, Christensen EI. Immunocytochemistry of renal membrane proteins on epoxy sections. Kidney International. 2007;**72**:731-735

[72] Ye L, Robertson MA, Hesselson D, Stainier DY, Anderson RM. Glucagon is essential for alpha cell transdifferentiation and beta cell neogenesis. Development. 2015;**142**:1407-1417

[73] Mojsov S, Heinrich G, Wilson IB, Ravazzola M, Orci L, Habener JF. Preproglucagon gene expression in pancreas and intestine diversifies at the level of post-translational processing. The Journal of Biological Chemistry. 1986;**261**:11880-11889

[74] Ravazzola M, Perrelet A, Unger RH, Orci L. Immunocytochemical characterization of secretory granule maturation in pancreatic A-cells. Endocrinology. 1984;**114**:481-485

[75] Varndell IM, Bishop AE, Sikri KL, Uttenthal LO, Bloom SR, Polak JM. Localization of glucagon-like peptide (GLP) immunoreactants in human gut and pancreas using light and electron microscopic immunocytochemistry. The Journal of Histochemistry and Cytochemistry. 1985;**33**:1080-1086

[76] Pelletier G. Identification of four cell types in the human endocrine pancreas by immunoelectron microscopy. Diabetes. 1977;**26**:749-756

[77] Braun M. The somatostatin receptor in human pancreatic b-cells. Vitamins and Hormones. 2014;**95**:165-193

[78] Cnop M, Hughes SJ, Igoillo-Esteve M, Hoppa MB, Sayyed F, Van de Laar L, et al. The long lifespan and low turnover of human islet beta cells estimated by mathematical modelling of lipofuscin accumulation. Diabetologia. 2010;**53**:321-330

[79] Borrett S, Hughes L. Reporting methods for processing and analysis of data from serial block face scanning

electron microscopy. Journal of Microscopy. 2016;**263**:3-9

[80] Denk W, Horstmann H. Serial block-face scanning electron microscopy to reconstruct three-dimensional tissue nanostructure. PLoS Biology. 2004;**2**:e329

[81] Heymann JAH, Gestmann M, Giannuzzi I, Lich LA, Subramaniam B. Site-specific 3D imaging of cells and tissues with a dual beam microscope. Journal of Structural Biology. 2006;**155**:63-73

[82] Kubota Y. New developments in electron microscopy for serial image acquisition of neuronal profiles. Microscopy (Oxford). 2015;**64**:27-36

[83] Kubota Y, Sohn J, Hatada S, Schurr M, Straehle J, Gour A, et al. A carbon nanotube tape for serial-section electron microscopy of brain ultrastructure. Nature Communications. 2018;**9**:437. DOI: 10.1038/s41467-017-02768-7

[84] Micheva KD, Smith SJ. Array tomography: A new tool for imaging the molecular architecture and ultrastructure of neural circuits. Neuron. 2007;**55**:25-36

[85] Peddie CJ, Collinson LM. Exploring the third dimension: Volume electron microscopy comes of age. Micron. 2014;**61**:9-19

[86] Ohno N, Katoh M, Saitoh Y, Saitoh S, Ohno S. Three-dimensional volume imaging with electron microscopy toward connectome. Microscopy (Oxford). 2015;**64**:17-26

[87] Pfeifer CR, Shomorony A, Aronova MA, Zhang G, Cai T, Xu H, et al. Quantitative analysis of mouse pancreatic islet architecture by serial block-face SEM. Journal of Structural Biology. 2015;**189**:44-52

[88] Nguyen HB, Thai TQ, Saitoh S, Wu B, Saitoh Y, Shimo S, et al. Conductive resins improve charging and resolution of acquired images in electron microscopic volume imaging. Scientific Reports. 2016;**6**:23721

[89] Ben-Othman N, Vieira A, Courtney M, Record F, Gjernes E, Avolio F, et al. Long-term GABA administration induces alpha cell-mediated beta-like cell neogenesis. Cell. 2017;**168**:73-85

[90] Bruneval P, Hinglais N, Alhenc-Gelas F, Tricottet V, Corvol P, Menard J, et al. Angiotensin I converting enzyme in human intestine and kidney. Ultrastructural immunohistochemical localization. Histochemistry. 1986;**85**:73-80

[91] Buffa R, Polak JM, Pearse AG, Solcia E, Grimelius L, Capella C. Identification of the intestinal cell storing gastric inhibitory peptide. Histochemistry. 1975;**43**:249-255

[92] Jonsson A, Ladenvall C, Ahluwalia TS, Kravic J, Krus U, Taneera J, et al. Effects of common genetic variants associated with type 2 diabetes and glycemic traits on α- and β-cell function and insulin action in humans. Diabetes. 2013;**62**:2978-2983

[93] Klöppel G, Lenzen S. Anatomy and physiology of the endocrine pancreas. In: Klöppel G, Heitz PU, editors. Pancreatic Pathology. London: Churchill Livingstone; 1984. pp. 133-153

[94] Suleiman H, Zhang L, Roth R, Heuser JE, Miner JH, Shaw AS, et al. Nanoscale protein architecture of the kidney glomerular basement membrane. eLife. 2013;**2**:e01149

[95] Wu M, Huang B, Graham M, Raimondi A, Heuser JE, Zhuang X, et al. Coupling between clathrin-dependent endocytic budding and F-BAR-dependent tubulation in a cell-free system. Nature Cell Biology. 2010;**12**:902-908

[96] Koga D, Kusumi S, Ushiki T, Watanabe T. Integrative method for three-dimensional imaging of the entire Golgi apparatus by combining thiamine pyrophosphatase cytochemistry and array tomography using backscattered electron-mode scanning electron microscopy. Biomedical Research. 2017;**38**:285-296

[97] Makhijani K, To TL, Ruiz-González R, Lafaye C, Royant A, Shu X. Precision optogenetic tool for selective single- and multiple-cell ablation in a live animal model system. Cell Chemical Biology. 2017;**24**:110-119

[98] Rizzo R, Parashuraman S, Luini A. Correlative video-light-electron microscopy: Development, impact and perspectives. Histochemistry and Cell Biology. 2014;**142**:133-138

[99] Yamashita S, Katsumata O, Okada Y. Establishment of a standardized post-embedding method for immunoelectron microscopy by applying heatinduced antigen retrieval. J Electron Microsc (Tokyo). 2009;**58**:267-279

Chapter 4

Transmission Electron Tomography: Intracellular Insight for the Future of Medicine

Abeer A. Abd El Samad

Abstract

Transmission electron microscopy (TEM) gives a good image for ultra-intracellular organelles in two-dimensional projections, but to get a three-dimensional structural information, we should use the high-voltage transmission electron tomography (TET). The use of TET is important to resolve the questions about the relation between the different cell organelles and their mechanisms of action to correlate structure to function for medical treatment solutions.

Keywords: transmission electron tomography, organelles, three-dimension, cells, histologists

1. Introduction

Transmission electron microscopy (TEM) had been largely responsible for shaping our views of organelle architecture, as it could provide the highest resolution within a spectrum of complementary tools used in the structural study of organelles in biological specimens. The images are two-dimensional projections, which pass through a relatively small slice of the specimen, and features from different levels are superimposed. This has enabled histologists in many circumstances to regard the third dimension as constant and interpret the image accordingly. In conventional thin-section TEM, the sections were generally much thinner than the specimen, and overlap of features was a problem. When the high voltage electron microscopes were used, the specimen with thick section-cut could be better examined. But the overlapped details of these images in two-dimensional images were more difficult to be interpreted. Transmission electron tomography (TET) resolved many of the limitations observed with serial thin section reconstruction by using sections thick enough (from 200 to 2000 nm) to contain a significant fraction of the organelle within the section volume allowing computing three-dimensional (3D) reconstructions of objects from their projections recorded at several angles by the use of high voltage (400–1000 kV) TEM [1].

Specimens were incrementally tilted in ET by a range up to ±60°, and many images were taken at each tilt. So, these serial images represented the whole specimen from different views. These serial images were aligned and then re-projected to give a 3D reconstruction or what is called a tomogram of the specimen. Therefore, the electron tomography represented the most available technique with high resolution to examine the biological specimens as cells [2, 3].

IntechOpen

2. Specimen preparation

The most common methods for preserving cell structure for TEM and TET were based on chemical fixation using glutaraldehyde for primary fixation and osmium tetroxide for secondary fixation. Following fixation, water was replaced by an organic solvent that was then replaced with a resin (**Figure 1**) [4].

Frozen-hydrated specimen was another method of preparation as a hydrated specimen was frozen rapidly enough, and then the liquid water was transformed into a vitreous solid state with a structure and density similar to those of the liquid. Sections of frozen-hydrated specimens were done using cryo-ultramicrotome and then examined by ET using a special holder. Cryo-sectioning combined the advantages of rapid freezing with the use of bulk material, and it was applicable to whole cells and tissues [1, 5].

Cryo-ET could directly image thin regions of cells that were adherent or grown on EM grids. So, it was becoming the method of choice for providing 3D information about intact intracellular structures at molecular resolution [6].

The plastic sections formed from resin cuts had more contrast because these sections were stained by heavy metal stains. Therefore, the images were been more close to focus. On contrast to the frozen-hydrated specimens, they had less contrast and should be imaged using a large defocus to get maximum phase of contrast [3].

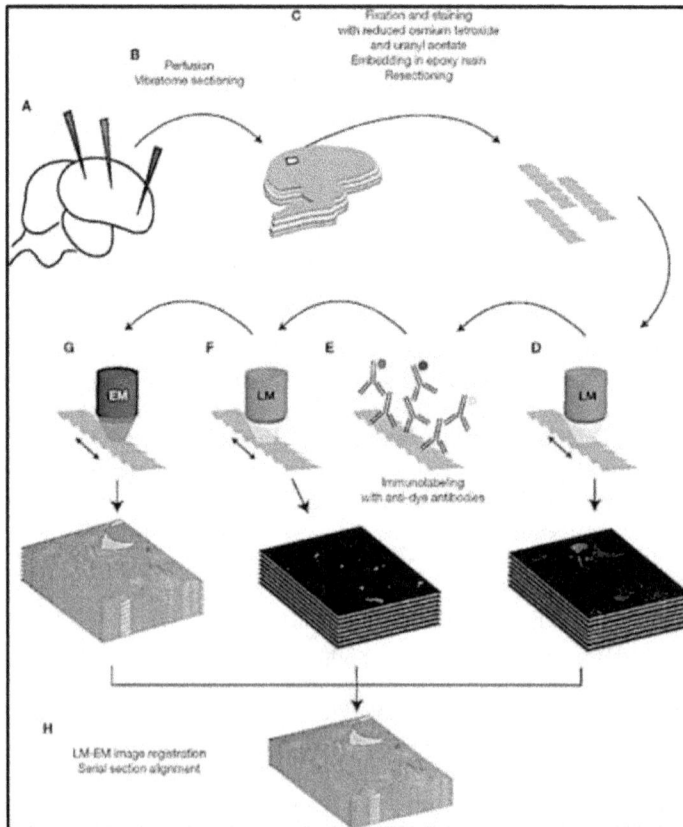

Figure 1.
Chemical fixation and plastic embedding [quoted from reference 5, available from: PubMed Central License: CC BY 4.0].

Sample thickness was generally limited to about 200 nm in 100 kV instruments and approximately twice than that in 300 kV instruments. While higher voltage EMs existed, these instruments were extremely rare and only moderately expanded (approximately another doubling) the permissible specimen thickness [3].

3. Reconstruction technique

Following image acquisition, image processing was divided into three phases. The first two phases were generally termed alignment and the last phase was the three-dimensional reconstruction proper. The images with tilt series should be processed to rotation and stretch to be aligned, and this was an important step to compensate any minimum difference in magnification, image rotation, and translation. The tilt axis of each image also was determined before the three-dimensional reconstruction [3]. Many software programs were used to perform these reconstructions as IMOD [7], TOM Toolbox [8], and SPIDER [9].

4. Examples of cellular organelles examined by TET to resolve questions about their mechanisms of action

4.1 Mitochondria

Mitochondria were among the first intracellular organelles to be studied extensively due to their importance in energy metabolism. Most observed mitochondria had large volume of matrix forming a compartment lined by inner membrane called the inner boundary membrane pushed against the outer membrane. A narrow space was seen between these two membranes of about 8 nm wide. The inner membrane projected into the matrix at discrete loci called crista junctions which were of a uniform diameter. The shape of cristae and the number of crista junctions were variable and dynamic in nature according to site of mitochondria and state of function. Electron tomography was also a valuable tool in showing the possible mechanisms of apoptosis as cytochrome c release was not effected by mitochondrial swelling and rupture of the outer membrane but apparently through the formation of large pores in the outer membrane through which intact cytochrome c could pass [1].

Using transmission electron tomography, investigators mapped the 3D topologies of some membrane organelles in the mouse ventricular myocardium, including transverse tubules (T-tubules), junctional sarcoplasmic reticulum (SR), and mitochondria. This research illustrated the geometric complexity of T-tubules. Electron-dense structures were usually seen between the outer membrane of the mitochondria and SR or T-tubules. Some investigators proposed that the relation between the mitochondria and the nearby structures are so important to local control of calcium in the heart, including the establishment of the quantal nature of SR calcium release, which is known as calcium sparks [10].

4.2 Endoplasmic reticulum (ER)

One of the benefits of using electron microscope tomography and live-cell imaging was to determine the mitotic assembly of the nuclear envelope, which was shown to be primarily originating from endoplasmic reticulum (ER) cisternae. Also, the nuclear pore complexes assembly occurred after the completely formed nuclear envelope. The chromatin-associated Nup107–160 complexes were in single units instead of assembled pre-pores. Therefore, the investigators proposed that the post-mitotic nuclear envelope assembled directly from ER cisternae which were followed by membrane-dependent formation of nuclear pore complexes [11].

4.3 Golgi apparatus

As the Golgi apparatus is so big and a single tomogram cannot capture much of its organization, "montage" tomograms were acquired from serial sections and then merged both laterally and "vertically" *in silico* to reconstruct a 4 mm^3 cellular volume. The 3D model of this merged serial tomogram illustrated the complexity of the true organization of the Golgi body and also led to the discovery of vesicle-filled "wells," which are formed by the aligned fenestrae from a series of cisternae [12].The inter-cisternal connections might function in retrograde and possibly anterograde trans-port of Golgi enzyme and lipids but not in transport of maturated components [13].

4.4 Desmosomes in cell junctions

A human skin biopsy sample was high-pressure-frozen, cryo-sectioned, and imaged by electron cryo-tomography to examine the desmosomes which are cadherin-mediated intercellular junctions. They are important to support the cell junction for tissue reinforce. The cryo-tomograms revealed that the cadherin molecules were densely and uniformly packed. They were at first arranged as small interacting groups with extracellular domains to form cis-homodimers then the opposing cell membrane approximated to form trans-homodimers. After the initial formation was established, more molecules were linked to the contact zone and the junction became more compacting. This process was regularized by build-ing blocks of alternate cis and trans dimers, and the strength of cell to cell contact became homogeneous. These processes were repeated to have finally a fully mature desmosome [14].

4.5 Actin filaments

By the use of cryo-ET, some investigators revealed that the membrane cytoskel-eton consisted of actin filaments mainly and the other associated proteins. This membrane cytoskeleton covered the whole cytoplasmic surface. Moreover, it was closely related to clathrin-coated pits and caveolae. The actin filaments which were linked to the cytoplasmic surface of the plasma membrane were likely to form the boundaries of the membrane compartments responsible for the temporary confine-ment of membrane molecules, thus partitioning the plasma membrane with regard to their lateral diffusion [15].

4.6 Microtubules, cilia, and flagella

The axoneme forms the essential and conserved core of cilia and flagella. Cryo-electron tomography was used to examine Chlamydomonas and sea urchin flagella to know information about the composition of axonemal doublet microtubules (DMs). These studies showed that B tubules of DMs contained 10 protofilaments (PFs) and also that the inner junction as well as the outer junction between the A and B tubules are different. The outer junction, important for the initiation of doublet formation, was formed by close interactions between the tubulin subunits of three PFs with unusual tubulin interfaces. The inner junction was formed of an axially periodic structure connecting tubulin PFs of the A and B tubules. The discovered microtubule inner proteins (MIPs) on the inside of the A and B tubules were observed to be more complex than previously thought, as they are composed

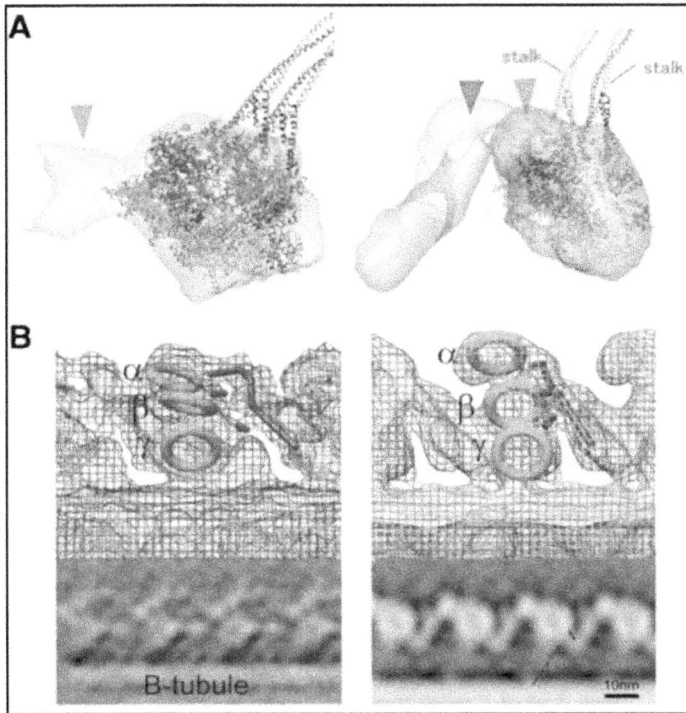

Figure 2.
Structural change of dynein induced by nucleotides: Left: structure of pre-power stroke; Right: structure without additional nucleotide (post-power stroke). (A) Tomography structure of mouse respiratory cilia consisting of two dyneins, the linker is shown in orange in pre-power stroke form and yellow in post-power stroke forms. (B) Tomography structure of Chlamydomonas, showing shift of the head as green rings and orientations of the stalk as blue and red dotted lines, as well as the neck domains and N-terminal tails (as red and blue solid lines) [quoted from reference 17, License: CC BY 4.0].

of alternating small and large subunits with periodicities of 16 and/or 48 nm. MIP3 formed arches connecting B tubule PFs. Also, the "beak" structures within the B tubules of the investigated Chlamydomonas DMT1, DMT5, and DMT6 were seen to be composed of a longitudinal repeating band of proteins with a periodicity of 16 nm [16].

The 3D structural analysis from cryo-EM has been playing indispensable role in motor protein research as a potential method to analyze 3D structure of complexes of motor and cytoskeletal proteins. The 3D image classification proved nucleotide-induced conformational change of dyneins and interesting distributions of multiple forms of dynein in the presence of nucleotides in cilia (**Figure 2**) [17].

Some authors reported that the centrosome-associated microtubule (MT) ends could be closed or open. The closed MT ends were more numerous and were distributed in a uniform matter around the centrosome. On the other hand, the open ends were found on kinetochore-attached MTs. These results showed the structural participations for models of microtubules interactions with centrosomes [18].

5. Telocyte; the entire cell examined by TET

A telocyte is a special type of interstitial cell having prolongations named telopodes [19, 20]. This cell has been described by electron microscope in different organs and tissues in the body. Some investigators examined the heart ultrastructure by TET and mentioned that telocytes make a network in the myocardial interstitium, which is involved in the long-distance intercellular signaling coordination. The cardiac telocyte network could integrate the overall "information" from vascular system (endothelial cells and pericytes), nervous system (Schwann cells), immune system (macrophages and mast cells), interstitium (fibroblasts and extracellular matrix), stem cells/progenitors, and working cardiomyocytes [21].

Other authors suggested that the telocytes present in the lamina propria of rat jejunum could be a heterogeneous population having different members which could switch between one activation states to another. They had cell to cell communication by paracrine mechanisms and to act as stem cell adjutants involved in epithelium renewal [22].

Transmission electron tomography has revealed complex junctional structures and tight junctions connecting pleural telocyte and small vesicles at this level in telopodes. Thus, pleural telocytes share significant similarities with telocytes described in other serosae. The extremely long thin telopodes and complex junctional structures that they form and the release of vesicles indicate the participation of telocytes in long-distance homo- or hetero-cellular communication [23].

6. Synapses in special sense organs

The cochlear inner hair cells (IHCs) are highly specialized cells in continuous stimulation and with rapid turn-over membrane. These cells synapse with multiple afferent nerve fibers. The ability of the IHC synapse to sustain activity for long periods is due to the presence of the synaptic ribbon. The use of 3D reconstruction by TET of an IHC infra-nuclear region revealed a network of rough endoplasmic reticulum (rER), mitochondria, and vesicles related to the synaptic ribbon (**Figure 3**). Small vesicles (about 36 nm) were seen by TET tethering to the rER. These vesicles were not observed either budding or fusing with the rER membrane, but only

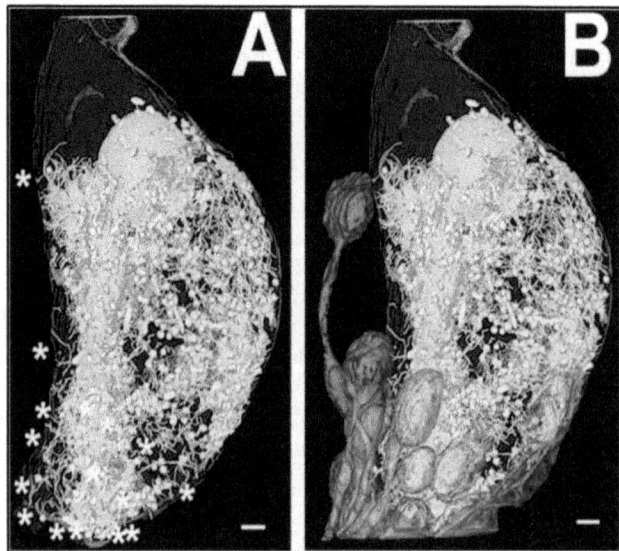

Figure 3.
Complete model of membranes, mitochondria, and synapse distribution in cell. (A) Ribbon synapses (red spheres) are marked with asterisks for clarity. (B) Reconstructions of synaptic terminals are shown (blue). [Quoted from reference 24, License: CC BY 3.0].

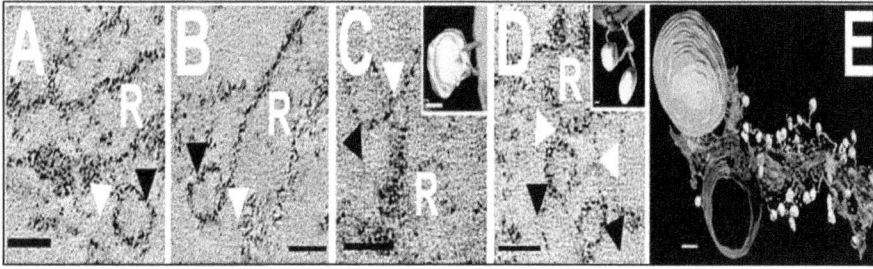

Figure 4.
TEM and TET reconstruction of vesicles link on rER and at the ribbon synapse. (A–D) Linkages (white arrowheads) are shown between membrane vesicles (black arrowheads) and three different areas of the rER (R). Insets in C and D show reconstruction of these links. (E) Reconstruction of a section of rER (green with ribosomes in red) with linkages to mitochondria (blue) and vesicles (yellow) showing the linkages surrounding the membrane. [Quoted from reference 24, License: CC BY 3.0].

connected to it by filamentous linkages. This result showed the possibility that this membrane network might represent a secondary store of neurotransmitter vesicles to be released during sustained synaptic transmission (**Figure 4**) [24].

7. Conclusion

Transmission electron tomography is not only an important instrument in the deep insight of the cellular components needed by histologists but also give information about mechanistic hypotheses, which may help scientists to correlate structure and function to different ways of getting diseased and to know the ways of medical treatment. It also helps to understand the disturbances of these organelles that cause the progression of damage in going more aged.

The integration between different basic sciences such as histology, pathology, physiology, bacteriology, and pharmacology, and the clinical physicians can be achieved through group of researches to cover some unclear points about different biological structures for the future of medicine.

Conflict of Interest

No conflicts declared.

Author details

Abeer A. Abd El Samad
Faculty of Medicine, Ain Shams University, Egypt

*Address all correspondence to: abirmohsen@yahoo.com

IntechOpen

© 2019 The Author(s). Licensee IntechOpen. This chapter is distributed under the terms of the Creative Commons Attribution License (http://creativecommons.org/licenses/by/3.0), which permits unrestricted use, distribution, and reproduction in any medium, provided the original work is properly cited. [cc] BY

References

[1] Frey TG, Perkins GA, Ellisman MH. Electron tomography of membrane-bound cellular organelles. Annual Review of Biophysics and Biomolecular Structure. 2006;**35**:199-224

[2] Frank J, Wagenknecht T, McEwen B, Marko M, Hsieh C, Mannella C. Three-dimensional imaging of biological complexity. Journal of Structural Biology. 2002;**138**:85-91

[3] Gan L, Jensen GJ. Electron tomography of cells. Quarterly Reviews of Biophysics. 2012;**45**:27-56

[4] Oberti D, Kirschmann MA, Hahnloser RHR. Projection neuron circuits resolved using correlative array tomography. Frontiers in Neuroscience. 2011;**5**:50

[5] Rigort A, Bäuerlein FJB, Villa E, Eibauer M, Laugks T, Baumeister W, et al. Focused ion beam micromachining of eukaryotic cells for cryoelectron tomography. Proceedings of the National Academy of Sciences of the United States of America. 2012;**109**(12):4449-4454

[6] Medalia O, Weber I, Frangakis AS, Nicastro D, Gerisch G, Baumeister W. Macromolecular architecture in eukaryotic cells visualized by cryoelectron tomography. Science. 2002;**298**:1209-1213

[7] Mastronade DN. Dual-axis tomography: An approach with alignment methods that preserve resolution. Journal of Structural Biology. 1997;**120**:343-352

[8] Nickell S, Forster F, Linaroudis A, Net WD, Beck F, Hegerl R, et al. TOM software toolbox: Acquisition and analysis for electron tomography. Journal of Structural Biology. 2005;**149**(3):227-234

[9] Shaikh TR, Gao H, Baxter WT, Asturias FJ, Boisset N, Leith A, et al. SPIDER image processing for single-particle reconstruction of biological macromolecules from electron micrographs. Nature. 2008;**3**:1941-1974

[10] Hayashi T, Martone ME, Yu Z, Thor A, Doi M, Holst MJ, et al. Three-dimensional electron microscopy reveals new details of membrane systems for Ca2+ signaling in the heart. Journal of Cell Science. 2009;**122**(Pt 7):1005-1013

[11] Lu L, Ladinsky MS, Kirchhausen T. Formation of the postmitotic nuclear envelope from extended ER cisternae precedes nuclear pore assembly. The Journal of Cell Biology. 2011;**194**:425-440

[12] Ladinsky MS, Mastronarde DN, Mcintosh JR, HowellO KE, Staehelin LA. Golgi structure in three dimensions: Functional insights from the normal rat kidney cell. Journal of Cell Biology. 1999;**144**:1135-1149

[13] Marsh BJ, Mastronarde DN, Buttle KF, Howell KE, McIntosh JR. Organellar relationships in the Golgi region of the pancreatic beta cell line, HIT-T15, visualized by high resolution electron tomography. Proceedings of the National Academy of Sciences of the United States of America. 2001;**98**:2399-2406

[14] AL-Amoudi A, Diez DC, Betts MJ, Frangakis AS. The molecular architecture of cadherins in native epidermal desmosomes. Nature. 2007;**450**:832-837

[15] Rigort A, Günther D, Hegerl R, Baum D, Weber B, Prohaska S, et al. Automated segmentation of

electron tomograms for a quantitative description of actin filament networks. Journal of Structural Biology. 2012;**177**:135-144

[16] Nicastroa D, Fua X, Heusera T, Tsoa A, Porterc ME, Linck RW. Cryo-electron tomography reveals conserved features of doublet microtubules in flagella. Proceedings of the National Academy of Sciences of the United States of America. 2011;**108**:E845-E853

[17] Ishikawa T. Cryo-electron tomography of motile cilia and flagella. Cilia. 2015;**4**:3

[18] O'Toole ET, McDonald KL, Mäntler J, McIntosh JR, Hyman AA, Müller-Reichert T. Morphologically distinct microtubule ends in the mitotic centrosome of Caenorhabditis elegans. The Journal of Cell Biology. 2003;**163**:451-456

[19] Popescu LM. The tandem: Telocytes–Stem cells. The International Journal of Biology and Biomedical Engineering. 2011;**5**:83-92

[20] Faussone-Pellegrini MS, Popescu LM. Telocytes. Biomolecular Concepts. 2011;**2**:481-489

[21] Gherghiceanu M, Popescu LM. Cardiac telocytes—Their junctions and functional implications. Cell and Tissue Research. 2012;**348**:265-279

[22] Crctoiu D, Cretoiu SM, Simionescu AA, Popescu LM. Telocytes, a distinct type of cell among the stromal cells present in the lamina propria of jejunum. Histology & Histopathology. 2012;**27**:1067-1078

[23] Hinescu ME, Gherghiceanu M, Suciu L, Popescu LM. Telocytes in pleura: Two- and three-dimensional imaging by transmission electron microscopy. Cell and Tissue Research. 2011;**343**:389-397

[24] Bullen A, West T, Moores C, Ashmore J, Fleck RA, MacLellan-Gibson K, et al. Association of intracellular and synaptic organization in cochlear inner hair cells revealed by 3D electron microscopy. Journal of Cell Science. 2015;**128**:2529-2540

Analytic Analyses of Human Tissues for the Presence of Asbestos and Talc

Ronald E. Gordon

Abstract

This chapter discusses the historic and current criteria for the analysis of cosmetic talcum powder and the finding the components of the talcum powder in human tissues. It describes how technicians and scientists have looked in the past for these components and how they should be looked at properly today. Within the chapter it has been shown that it can be complicated, especially when the tools and the methods used are not adequate or sensitive enough. It also goes on to describe methods for analysis that are sensitive enough in both mineral analyses and in human tissue. It also defines the terms that are necessary to use for inclusion of structures based on the scientific knowledge we have today not confused with what either industry or their defenders are trying to use to confuse or defend their positions.

Keywords: electron microscopy, human tissue talc components

1. Introduction

One of the best and concise reviews of what has been defined as asbestos is in a report by the U.S. Department of the Interior: U.S. Geological Survey by Virta [1]. Briefly, asbestos and talc are minerals that are mined from the earth. The asbestos is defined as having six different types of magnesium (Mg) silicates (Si). An important feature of many of the types is that they may or may not contain iron when they are removed during mining. They can have other ions as well and those include sodium (Na); calcium (Ca) or manganese (Mn) along with the Mg and Si. These minerals are defined by the presence of these elements and their ratio one to another. There are numerous publications defining the mineralogic nature of these minerals throughout the literature besides what is stated in Virta [1]. They are also defined initially by their color when in the ground and raw, by their size, shape, how they were formed and their crystalline structure by light and electron microscopy. The types are divided in two groups, serpentine and amphiboles. The serpentines principally are chrysotile and the amphiboles consist of five different types based on the ratio of the Mg to Si and other elements that are integrated into the molecular structure. The amphiboles consist of crocidolite, amosite, anthophyllite, tremolite and actinolite. They are mined in both open and closed type mines. The elemental composition of these six types of asbestos is seen in **Table 1**. The talc is a basic $H_2Mg_3Si_4O_{10}(OH)_2$.

These mined minerals, both asbestos and talc, have in the past been used in many products and have been shown to have detrimental effects in humans and animals when they enter the body of these organisms [2]. The effects range from tumors to fibrosis. The asbestos has been classified as a carcinogen and has been known to cause lung cancers, mesotheliomas, gastrointestinal cancers and more recently has been implicated in causing ovarian cancers and others. In the case of talc and talcum powder products, it has been implicated as causing these tumors either indirectly because it is contaminated with asbestos or the talc is a carcinogen or co-carcinogen even without the contaminating asbestos. Further, asbestos, generally in high doses is a well-known cause of interstitial fibrosis, asbestosis, and pleural plaques in lungs. Talc in relatively high doses is also known to cause fibrotic lesions, specifically in the lung. This type of fibrosis is referred to as granulomas.

The mechanisms of causation of these diseases have been shown to be either direct or indirect. What I mean by direct is the interaction of the asbestos fiber or talc fiber or particle with DNA in the cell eliciting mutations. The indirect methods of causing these same mutations is the release of oxidants either within the cells or from macrophages that have either completely engulfed the fibers or particles or partially engulfed them because they are just too large to be contained within the cells. These oxidants cause DNA mutation, which can cause the cells to convert to cancer cells. In addition, when these fibers and particles get into the cell, the cells are known to release cytokines and chemokines that can result in the recruitment of inflammatory cells and the in the development of the fibrotic lesions.

The relationship between asbestos fibers and amounts, size, dimensions, and type has been correlated with the development of diseases. It has been determined that the greater amount of asbestos present in the peripheral lungs or other tissues of known tumorigenesis, the greater the risk of developing that tumor or fibrotic change. The longer and thinner the fiber, the greater the risk. Also, amphiboles bare a greater risk than chrysotile unless the chrysotile is relatively long fiber type and numerous and even than it must require a greater latency between exposure and the development of tumors specifically. Crocidolite by far is considered the most carcinogenic with amosite not far behind and then anthophyllite. Tremolite and actinolite tend to parallel chrysotile because they are generally shorter and less numerous because they are contaminates with the chrysotile or talc.

An important link and correlation between environmental exposure and causation of the diseases describes above is the finding of these particles in human tissue. The remainder of this chapter is dedicated to the specific criteria and methodologies for defining and identifying these fibers and particles in human tissues which can be very different and much more difficult to identify in from those evaluated from the same minerals that come directly from mining. This chapter addresses many of these issues in defining these fibers after they have been subjected to tissue modification after entering the human body.

NAME	COMPOSITION
CHRYSOTILE	$(Mg_6O_4(OH)_8)^{-4}$ ($Si_4O_{10})^{-4}$
RIEBECKITE (CROCIDOLITE)	$Na_2(Fe^{+2\cdot}Mg)_3Fe^{+3}Si_8O_{22}(OH)_2$
GRUNERITE (AMOSITE)	$(Fe^{+2})_2(Fe^{+2},Mg)_5Si_8O_{22}(OH)_2$
ANTHOPHYLLITE	$Mg_7Si_8O_{22}(OH)_2$
TREMOLITE	$Ca_2(Mg_5Si_8O_{22}(OH)_2$
ACTINOLITE	$Ca_2(Mg,Fe^{+2})Si_8O_{22}(OH)_2$

Table 1.
This table illustrates the chemical composition of the various asbestos fibers.

Analysis of human tissues for the presence of asbestos and talc is nothing new [3, 4]. However, what makes this type of analysis unique and now very much at the forefront is that these components have been identified in and possibly attributed to the development of tumors not considered in the past [3].This author has looked at numerous types of tissues and tumor tissue from same and in a variety of other tissues and organs [5–7]. What is emerging is the question of protocols and methods for detecting these particles and fibers, identifying them and attributing them to the disease processes. The history behind asbestos exposure and disease is well documented with causing lung tumors, mesotheliomas in pleura and abdomen and asbestosis in lung. Only some of the effects of talc have been documented, most of which are associated with the development of granulomas in the lung and in the pleural spaces when the talc is injected into the space to avoid the accumulation of fluid, talc pleurodesis.

More recently, within the last 10 years there has been attribution of cosmetic talcum powders causing mesotheliomas and possibly other lung tumors [4]. However, the attribution has been directed to the contaminating asbestos in the product [4]. Companies that currently sell and those that sold this product in the past are claiming that their products are free of asbestos. They base this on tests that have been done in a number of laboratories using a variety of testing protocols. However, further testing using more sensitive methodologies have shown these products to still contain asbestos.

It is the specific intent of this chapter to address all the issues with regard to the methodology of testing of cosmetic talcum powders for the presence of asbestos and to be able to document the presence and type of asbestos in human tissue studies in persons that have used these products with no history of exposure to asbestos from other sources and differentiate the particle type.

2. Historic testing of cosmetic talcum powders

The testing of talcum powder goes back to 1968, Cralley et al. [8] tested 22 different samples of talcum powder off the store shelf for fibrous and mineral content. They found that all 22 containers had a significant amount of fibrous components by light microscopy and phase contrast (PCM). The type of fibers were not identified by PCM or by XRD and assumed to be fibrous talc with smaller contaminates of tremolite, anthophyllite, chrysotile and pyrophyllite. Without identifying the fiber types, they identified fibers that could not be seen by light microscopy and concluded that it was these fibers that could be the source for ferruginous bodies seen in humans.

In the 1970s, numerous investigators analyzed talcum powders. Walter C. McCrone Associates, Inc. looked at talcum powders for a variety of different companies and groups including NIOSH. They used polarized light microscopy (PLM), XRD and Transmission Electron Microscopy (TEM) and reported finding asbestos fibers in many of the samples [9–12]. In 1972, at New York University Chemistry Department tested a sample of a specific talcum powder called 1615 [11]. XRD indicated that the fibers were suspect for asbestos and then the talc was subjected to a more critical testing where they identified both tremolite and chrysotile [13].

In 1974, Rohl and Langer [14] tested a number of talcum powder specimens using both light microscopic techniques, XRD and analytic electron microscopy (ATEM) with selected area electron diffraction (SAED) and electron microprobe and indicated they were able to detect only a very small amount of the fibrous asbestos particles by PLM or XRD mainly because of the size of the particles and recommended that it was essential to do analytic TEM to analyze talcum powder for the presence of asbestos. In another article in 1974, Rohl [15] indicated that the asbestos in the talcum powder was directly from its mining. However, in both studies they concluded that a negative finding of asbestos in these products by just XRD could mean there were possibly billions of fibers in just a half gram of the talc if tested by a more sensitive technique, i.e. TEM.

In 1976, Rohl and Langer [16] reported on 20 off the shelf talc or talcum powder products of which they were able to detect asbestiform fibers in 10 of the 20. They used a combination of XRD, PLM, scanning electron microscopy (SEM) and TEM. They used EDS and SAED with the TEM samples to identify the asbestos fibers. They concluded that the great majority of the talc asbestos fibers tested by XRD and PLM would go undetected as compared to SEM and TEM specimens.

In 1990, Kremer and Millette [17] published on the same powder used by the McCrone Laboratory in 1985 and employed a different methodology of suspending the material in a solution of methylcellulose to view the fibers by TEM and found a variety of different minerals, including asbestos.

3. Historic methods for observing talc components

What is of great interest is that there are two methods promoted by the cosmetic industry, CTFA-J4-1 [18] and USP-Talc [19] which only employ XRD and light microscopic techniques. They also state that using TEM with SAED is much more sensitive technique but they do not recommend using that methodology. The unfortunate part of all this is that the industry relies on this method of testing knowing full well that they will not find the great majority of contaminating asbestos fibers by these techniques.

However, there have been many techniques published that use a combination of both XRD, light microscopy (PLM or PCM) and which state that if there is a negative finding by these techniques it is important to look by TEM or use SEM as a screening technique. Some of these techniques include the EPA 1993 bulk method [20] as one such method. Most of the techniques not only require that TEM be used, but both SAED and EDS be performed to be able to determine the identity of the type of fiber one is seeing. These include the AHERA methodology which employs the Yamate et al. [21] method. Other methods that are frequently used are those from ASTM D6281 [22], D5755 [23], D5756 [24], and D6480 [25] all of which require TEM. There are two others called ISO10312 [26] and ISO13794 [27] which are very much the same as the ASTM methods. The techniques for verification of asbestos fiber types require SAED confirmation. However, in some cases where there may potentially be a question or a problem of confirmation zone-axis maybe required and is described in both the ASTM D6281 [28] technique and in Yamate et al. [29]. However, this is only if there is a question, since in most instances it does not give further support to routine SAED. When combining the newer more sensitive EDS equipment with SAED, zone-axis analysis will not add anything. The most important point here is that there is no specific defined method for identifying asbestos in talcum powder products. However, the use of the most sensitive techniques available is imperative.

4. Differentiating asbestos fibers

There are six types of asbestos that have been described and identified as detrimental. These are categorized into two types, serpentine, chrysotile, or the amphiboles, crocidolite; amosite; anthophyllite, tremolite and actinolite. As seen in Chart 1, the six asbestos types and talc show their chemical composition and how they are both very similar and or different based purely on their chemical composition. In differentiating asbestos fibers there are two approaches to the problem that can/are seen differently depending on who is looking at the fiber(s). This includes a definition of what asbestiform means and based on who is looking will determine which definition may be applied. When a mineralogist is looking at fibers their criteria requires a population of fibers that have to meet the 3:1 ratio, equal to or longer than 5 μm with parallel sides that has grown in an asbestiform mineral habit. On the other hand, when viewed by someone looking at single fibers the only distinctions that can be made are based on the observed morphology of that fiber. It is not possible to relate it to the environment from which it was formed. The criteria under the latter situation is the one that all government agencies adhere to and require for it to be an asbestos fiber and that is that the fiber should be greater than 0.5 μm in length, have at least a 3:1 ratio of length to width and have parallel sides. That is what qualifies it to be an asbestos fiber.

In 1990, Wylie [30] published some suggested criteria which were primarily based on light microscopic criteria and not electron microscopy. Wylie et al. [31] suggested that it had to have a 20:1 or greater and had to be very thin fibers or fibrils, less than 0.4 μm in width and two other criteria which included parallel fibers in bundles, splayed ends of fiber bundles, fibers in the form of thin needles, matted masses of individual fibers and finally fibers showing curvature to be considered as asbestos. In the EPA R-93 [32] this was repeated in the glossary. However, it is possible to see that these light microscopic criteria are useless when viewing a single fiber or fibril by transmission electron microscopy. It has been determined that if one were to use this criteria, approximately 80% of the asbestos fibers would be misclassified. EPA R-93 method [32] suggest the use of 10:1, ratio based to some degree, on a 1985 Wylie publication [31] indicating that if 20:1 were used with an amosite population, as much as 50% of asbestiform asbestos fibers would not be counted. Even the bureau of Mines Circular [32] indicates that a 5:1 ratio is the most realistic. The 5:1 ratio is in fact used by AHERA, ASTM methods D6281, D5755, D5756 and D6480 and ISO 10312 and 13794. The width of the fiber as described by Harper et al. [33] seems to be the best discriminator. In a publication by Kelse and Thompson [34] from RT Vanderbilt further supports the concept that any fibers equal to or greater than 5 μm in length and less than 0.25 μm in diameter are asbestos fibers and almost all less than 0.5 μm in width are fibers and not cleavage fragments. However, these are purely mineralogy distinctions and have virtually no application to biologic systems since the cells that are activated by these fibers in human body do not make these distinctions. The cells only are effected by the shape, size dimensions and surface charge on these fibers which can cause a form of oxidant injury or mechanical alteration of the cellular DNA in the mesothelial or ovarian epithelial cells that take them up and in macrophages and inflammatory cells that engulf them causing the release of cytokines, chemokines and molecules associated with oxidant injury which can indirectly effect mesothelium and ovarian epithelium to become tumors. Of course this excludes the concept that the same molecules can also cause the development of fibrosis or asbestosis. Therefore, this entire argument rose by a very few mineralogists that cleavage fragments not be considered as harmful, is just wrong.

The other issue that arises from a similar argument is talc itself. Talc can also be present in the form of fibers that can mimic, but can be differentiated analytically from asbestos and can cause fibrotic lesions in some mammals and in human lungs [35, 36]. Therefore, it is realistic to consider talc, especially in the fibrous form, a potential causative factor in the development of mesotheliomas and ovarian cancers. There is, has been, and is currently significant research ongoing to prove that the talc can be considered a carcinogen, alone, as a co-carcinogen with the asbestos or as a promoter with the asbestos, just based on its ability to produce an inflammatory response.

Zone indexing of asbestos fibers and talc fibers for the purpose of differentiating them has been described and shown to be relatively unnecessary procedure [37]. EDS spectra can be indistinguishable between anthophyllite and talc [37]. However, when anthophyllite is compared to talc fibers by SAED talc fibers no matter how they are turned or tilted show the typical hexagonal pattern. On the other hand, anthophyllite can only show a pseudohexagonal pattern if tilted to a specific angle. Therefore, the only issue would be that one would see less anthophyllite if tilted in that specific angle as compared to talc, but talc would never be confused with anthophyllite if SAED is performed in only a single angle.

5. Analysis of human tissues

There have now been many reports, possibly hundreds that describe the protocols for identifying asbestos and talc in human tissues. However, when one looks at these protocols it is possible to break them down to three similar, but yet different means of looking for these particles in these human tissue preparations. As fully described below, the remaining material after tissue digestion can be prepared by the filtering method a portion of the filter is put directly on an SEM stub and then analyzed. The Alternatively small portions of the filter can be placed onto TEM grids and then observed by either SEM or TEM. Lastly, the material can be placed directly on a formvar support film on a TEM grid and then directly analyzed by TEM.

6. Analysis of asbestos by SEM

There are at least two investigators that look at human tissue preparations; one of which has been doing these analyses for years by SEM and that is Roggli [38]. Based on all government criteria SEM analysis is not an acceptable criteria. All government agencies that describe doing electron microscopy observation and identification of asbestos require TEM with at the least SAED, but EDS is always listed as a criteria. SEM analysis does not allow the technician, examiner or scientist to evaluate the crystalline structure of the fiber or particle of interest. SAED is what is considered the gold standard for identifying asbestos fibers and other particles such as talc. As will be shown as this explanation unfolds, the identification of asbestos fibers and specific types of asbestos fibers in human tissues is far more difficult than that of the mineralogist identifying them from ground up rocks or mined minerals. This is the case mainly because the longer these fibers are present in a biologic environment with cells, tissues, animals or humans, the fibers are modified and frequently can only be distinguished using SAED. When SEM is used there is a significant potential for error. The error is most likely to occur when distinguishing fibers between anthophyllite, chrysotile, tremolite and non-asbestos talc fibers. Pure morphology by SEM on single fibers is very similar in appearance. EDS analysis of the same fibers are more difficult to get the optimal elemental composition because the electron beam energy is significantly lower, generally never more than about 40 KV whereas in a TEM it is generally 75–200 KV. It has been long known that the higher the KV the greater the penetration of the beam into the fiber. Lower energy levels will only affect the very surface of the crystalline structure or fiber. As stated before, fibers removed from biologic systems are modified as their surfaces by the interacting environment. Biologic interactions results in the removal of molecular components from the surface referred to as leaching. The leaching is mostly associated with removal of magnesium, which can lead to the change in the Mg to silica, Si ratio which can put fibers into different categories or types based purely on elemental analysis. The most effected fibers or particles are chrysotile type asbestos fibers and talc fibers and particles. These are most susceptible to leaching and ultimate relatively rapid breakdown of the structure. Examples will be given below when discussing changes in TEM. The alternative to leaching is that elements in the form of molecules can become adherent to the fiber or particles. The most common element and ion that adhere is iron, Fe. When the fibers or the particles are present in tissue for long periods, years, the iron, in combination with protein molecules can produce ferruginous bodies or asbestos bodies on asbestos fibers. When there are substantial amounts of iron and protein to form bodies they are easy to identify even by light microscopy. However, there can be lighter coatings not forming the pearl like structures on the fiber or covering the particles and then it is just seen as increased iron which could lead to an inaccurate identification by EDS analysis which has already been argued in letters to the editor following a publication [2] where one laboratory wanted to identify an anthophyllite fiber as an amosite fiber. In addition, other elements such as sodium, Na, aluminum, Al and calcium, Ca, can adhere to the fiber surface also leading to a misidentification when looking at fibers with the SEM by morphology and EDS alone. This will be discussed later with examples in the TEM section.

7. Methods for SEM or ATEM preparation

7.1 Methods for filtering

The filtering methodology has been published many times and is used by laboratories that evaluate air, water, bulk and human tissue samples [29]. With human tissue samples the material must first be digested and cleaned with distilled water to remove any biologic material. This is performed by a variety of techniques which have previously been employed. When the tissue is received in formalin, the tissue is either dried or completely and weighed or is just blotted dry. In the former the results will be expressed as dry weight and the later wet weight. Either way they are approximately comparable by approximately a factor of 10. Either way the tissue is then treated with either hypochlorous acid, (Clorox) or 5% potassium hydroxide, KOH, which acts to digest away any biologic material or it becomes soluble in either solution. The inorganic material is then separated by centrifugation and repeated, ×5, sequential washes in distilled water. The remaining inorganic and metal materials are then put into a final suspension of distilled water and filtered onto either polycarbonate or missed ester type filters. After drying the filters are cut into small pieces and placed on formvar coated copper or nickel locator grids or directly onto a SEM stub. The filters are lightly coated with evaporated carbon to help prevent transposition or release of the fibers and particles during the collapse protocols. The filters are collapsed with either acetone or ether depending on filter type. Some investigators use low temperature ashing to remove any residual biologic material, however that is rarely done today. The ashing was most often used for filters that were prepared from water and air sampling which where the material present on the filters is not predigested with Clorox or KOH. The grids or stubs are then ready for observation.

7.2 Methods for drop method

An alternative method that this author has used for over 45 years was first defined by Langer et al. [39] where the digested material is resuspended in a known amount of distilled water and then 10 μl drops are placed directly on formvar coated grids and dried. The grids are then ready to view by scanning or transmission electron microscopy evaluation.

8. Methods for asbestos fiber and talc particle analysis

The problems associated with SEM have been defined above and will not be discussed here.

When we look at the prepared grids, whether the grids are viewed and evaluated by three criteria, morphology, EDS and SAED, the grids are first scanned to make sure that they have less than 5% broken openings. Dependent upon the criteria used in the laboratory, the grids are critically evaluated at magnifications between 10 and 20 K, one grid at a time, for the presence of asbestos fibers or whatever is being evaluated.

To determine if a fiber is asbestos is based on well-established criteria. If a fiber has parallel sides and has a 3:1 or 5:1 aspect ratio it has the morphological criteria for a fiber. If the fiber demonstrates individual smaller components within the larger fiber, referred to as fibrils, each of which is a fiber if seen alone is than a better criteria for the classification as asbestiform by mineralogy criteria. When pathologically evaluating the morphology of an asbestos fiber a mineralogists criteria of being asbestiform or grown in an asbestiform habit is not at all considered. However, when the asbestos fibers are seen as a bundle, would be considered an asbestos fiber by either a pathologist or mineralogist. The chemistry of the fiber has to contain specific elements which include sodium, magnesium, silica, calcium, manganese and iron. They also have been found in approximate ratios using silica as a reference. Each type of asbestos type has a specific ratio when in its natural form. Crocidolite, amosite and tremolite are generally easily recognized by EDS alone. Chrysotile and anthophyllite can look very similar. Also one must exclude fibrous talc when making these determinations since it too looks very much like chrysotile and anthophyllite by EDS alone. The third criterion is selected area electron diffraction which determines crystalline structure of the fiber or material. This technique produces patterns that identify the crystal very much the way fingerprints identify people. When a fiber cannot be identified by morphology and EDS, SAED is the determining technique. SAED can only be performed with a transmission electron microscope. It is possible to screen for fibers and particles by XRD, PLM, PCM and SEM, however, for definitive identification, TEM using morphology and SAED or ATEM using both EDS and SAED are absolutely required. Even then it may be difficult to identify the fiber type because of all the issues described above as interference in the ability to specifically identify a fiber. Amosite and crocidolite are generally the easiest to identify. Chrysotile, anthophyllite and fibrous talc can easily be misidentified. Tremolite/ actinolite can also be determined but with difficulty and the use of SAED to differentiate it from chrysotile, anthophyllite or fibrous talc. So when evaluating human tissue isolation of fibers and particles, there are many elements present in tissue that can ionically or covalently adhere to the outer most part of the fiber or particles. A few of these elements include sodium (Na), aluminum (Al), calcium (Ca) and iron (Fe). When it is not possible to identify any features by morphology, most laboratories first focus on the EDS which will give us the element composition and the ratio of one element to another. One has to consider that the fibers and particles once in a cell are attached by acids, and enzymes that can modify the surfaces by eroding the fiber, usually by leaching the Mg. However, many elements can be added to the surface. When Na is added and Mg is partially leached amosite can appear to be crocidolite (**Figure 1**). When Mg is leached from anthophyllite and Fe is added it can appear to be amosite (**Figure 2**). When chrysotile has Mg leached and Fe added in appears to be anthophyllite (**Figure 3**). It is very difficult to sort between tremolite and actinolite because Fe can be added. Fibrous talc can look like chrysotile and anthophyllite by EDS only (**Figure 4**). If there is a lot of Calcium phosphate as background and interference with some added iron, it may not be possible to confirm tremolite or actinolite by EDS (**Figure 5**). One more exhibits anthophyllite with leached magnesium and some added iron; however, it could be easily be confused with being tremolite or actinolite (**Figure 6**). There are many cases like this that end up being defined by the SAED and not EDS and morphology alone.

The series of fiber TEM micrographs and their corresponding EDS show how difficult it could be with only morphology and EDS to define asbestos fiber type removed from humans. It then becomes critical to perform SAED on these fibers to determine the crystalline structure based on the dispersion patterns. However, this technique can also be problematic in identifying fiber types. It is possible if the anthophyllite is tilted just right it can look like talc in the SAED pattern and if the D-space measurement can also be the same [29, 37]. However, the opposite is not true. Talc never looks like anthophyllite by SAED pattern or d-spacing measurements. As a result the only effect this could have is to reduce the amount of anthophyllite if present.

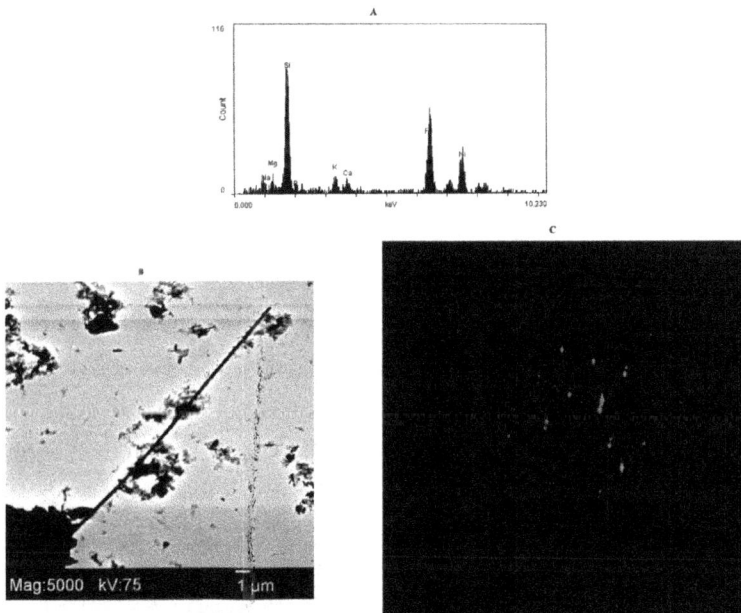

Figure 1.
This EDS spectrum (A) represents the asbestos fiber seen in (B). At first look this long narrow fiber would correlate with the spectra of a crocidolite fiber having approximate ratios of 1:1:10:6 Na:Mg:Si:Fe. The potassium, K, is from adherent from the tissue digestion and the calcium, Ca, is surrounding interference material with the phosphate, P. However, when SAED was performed, the diffraction pattern was that for amosite and not crocidolite indicating that the sodium, Na, was either interference from the surrounding area or adherent to the fiber itself. The SAED in 1C confirms that it is amosite.

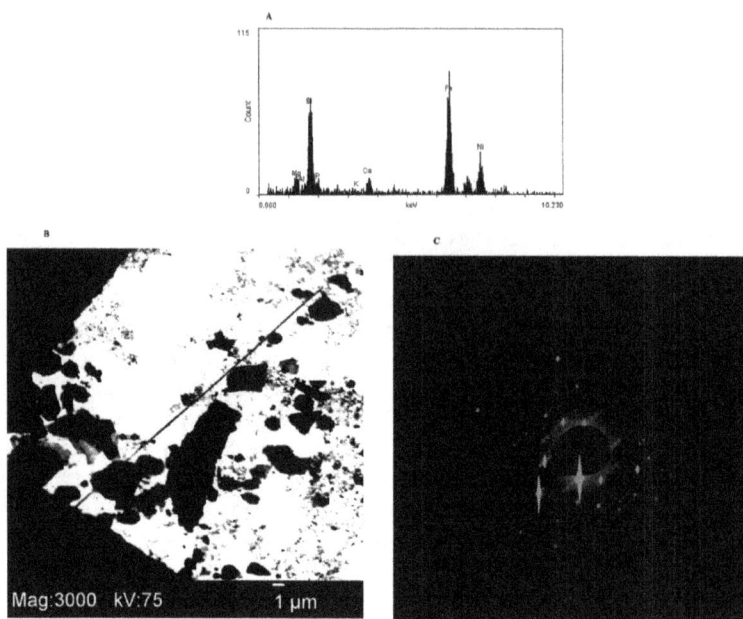

Figure 2.
This EDS spectrum (A) represents the asbestos fiber seen in (B). Other than the calcium phosphate, CaPO₄, it would be consistent with it identifying an amosite asbestos fiber. However, there is more Fe than would be expected and a little less Mg. When SAED was performed, this fiber turned out to be anthophyllite with significant Fe more than likely coming from interference iron particles surrounding the fiber and some leached Mg. The SAED in 2C confirms that it is amosite.

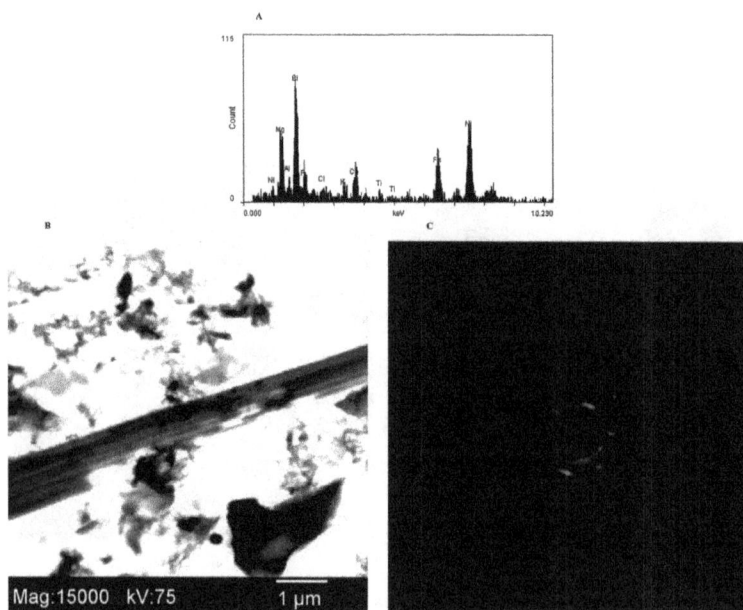

Figure 3.
This EDS spectrum (A) represents the asbestos fiber seen in (B). This EDS would best fit anthophyllite type asbestos with some increased Fe. There may be a slight too much Mg for anthophyllite. SAED of this fiber proved to be chrysotile with increased Fe and Mg leaching. The other elements identified, calcium phosphate, potassium chlorine and a little aluminum are from the surrounding interference. The SAED in 3C confirms that it is chrysotile.

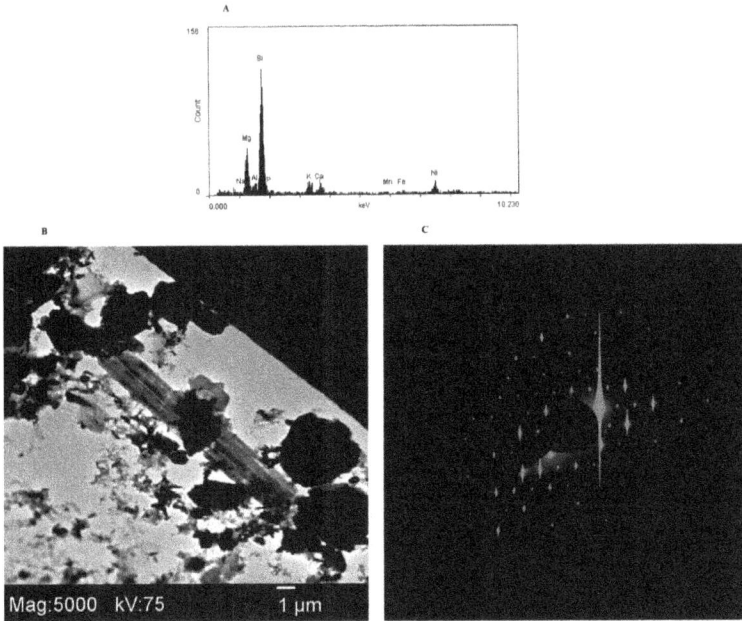

Figure 4.
This EDS spectrum (A) represents the asbestos fiber seen in (B). This EDS could represent a chrysotile asbestos fiber with leached Mg or an anthophyllite type fiber with no Fe. However, SAED exhibited the classic hexagonal pattern of a talc fiber. The SAED in 4C confirms that it is talc.

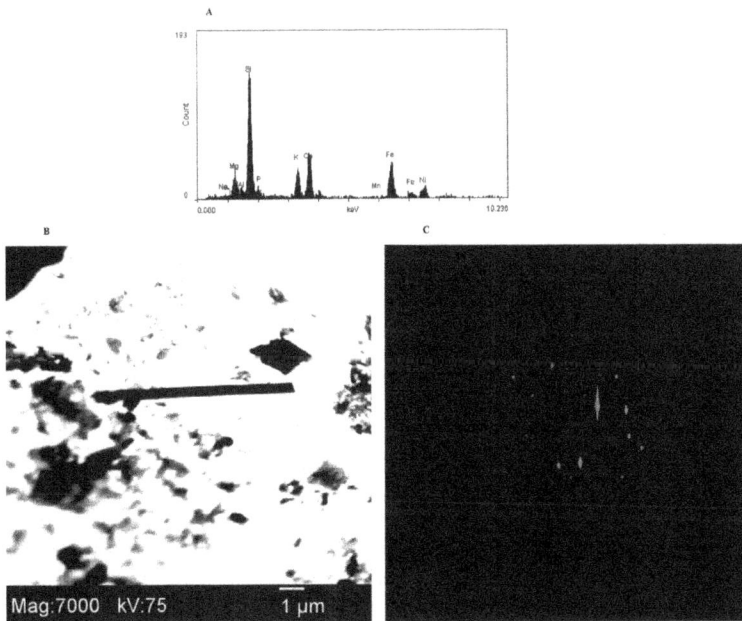

Figure 5.
This EDS spectrum (A) represents the asbestos fiber seen in (B). This appears to be anthophyllite asbestos with some leached Mg and slightly more Fe. The K is from the digestion and there is some calcium phosphate. SAED exhibits the typical pattern for tremolite/actinolite type asbestos. This further demonstrates that the much of the calcium was from the fiber and not the surrounding calcium phosphate. The SAED in 5C confirms that it is tremolite.

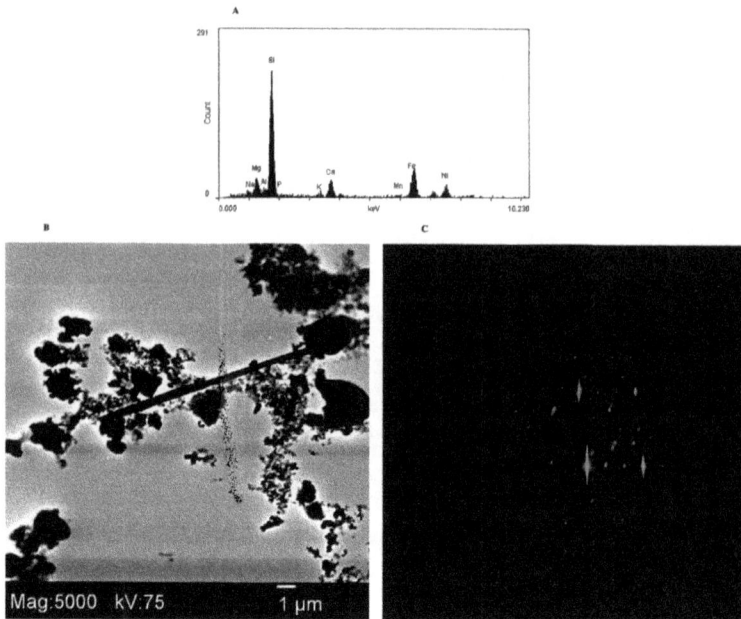

Figure 6.
*This EDS spectrum (A) represents the asbestos fiber seen in (B). This EDS appears to represent a tremolite/
actinolite fiber. There is a little added Fe. However this is an anthophyllite asbestos fiber based on the
SAED. There is interfering Ca particles, in this case it was not associated with phosphate. The SAED in 6C
confirms that it is anthophyllite.*

One other issue that must be addressed is that of cleavage fragments in human
tissue. The entire concept of determining if a fiber is asbestos or a cleavage fragments by mineralogists is defined by whether it is asbestiform or not. By their definition asbestiform relates to the way that the manner in which the crystals initial
formed. They would refer to it as asbestos only if it was formed in an asbestiform
habit, meaning that all the fibers were completely linear and just seen together as a
bundle very much like a thick telephone wire. Any other type of arrangement would
be considered a cleavage fragment as it may separate from the larger mass. If that
larger mass was not asbestiform and it was possible to see a structure that resembled
a fiber with parallel sides, the mineralogist would call it a cleavage fragment based
on knowing that none of the particles in the larger population were linear fiber
types, as they refer to them as asbestiform. It has been shown that the great majority
of fibers that may be considered to be cleavage fragments are generally very short
with very small aspect ratios. They most often look like chunks rather than fibers
and are also generally thicker than fibers seen as asbestiform asbestos fibers. There
have been papers published that indicate that if the fiber has a 20:1 ratio then it is
asbestos [30]. There are also papers that indicate that a ratio of 8:1, they are asbestos
[33]. Possibly the best criteria is when a fiber has a minimum aspect ratio of 5:1 and
the width is 0.25 µm or less it is definitively asbestos [34]. The definition of asbestiform for a pathologist or appropriate testing laboratory or someone looking for
asbestos fibers in tissue is purely based on the criteria of having a length to width
ratio of 3:1 or 5:1 with parallel sides. In the absence of a population of fibers and the
mineralogical identifiable non-asbestiform mineral there are no reliable criteria at
the light or electron microscopic level to call it an asbestos fiber or a cleavage fragment other than those relating to size and shape described above.

There are two basic reasons that the above criteria for fibers are not considered cleavage fragments, but asbestos fibers, if from human tissue. The first of which is when such a fiber is analyzed from a human tissues preparation and there are only a few fibers it is impossible to identify it as from an asbestiform habit of growth. Therefore, all governmental organizations only refer to the criteria of aspect ratio and parallel sides. Further, it has been determined that the fibers identified as asbestos, the size, shape and type of fiber is critical in attributing it to causation. The longer and thinner fibers have been most commonly attributed to tumor and asbestosis development. Another criteria is that the charge distribution on the surface of fibers, asbestos by mineralogy definition or cleavage fragments are not significantly different from asbestiform fibers defined by mineralogist's criteria and will have the same oxidative effects which indirectly cause genetic or DNA mutations or elicit chemokines or cytokines resulting fibrosis, asbestosis. Lastly, fibers identified as asbestos found in human tissue analyses have been attributed to and correlated with the history of exposure and the above diseases in tens of thousands of cases over the last 60 years. This is without determining if it is a cleavage fragment or not.

9. Background controls

Background controls are imperative when performing asbestos fiber burden analyses on human tissues. It is extremely important because without them there is no criteria for comparison to assess whether what a scientist or technician is finding has any relevance with regard to exposure history and what remains in the tissue depending on latency or how long it has been since the patients' exposure(s).

To attribute a patient as a background exposure, it is imperative that a complete patient history must be taken by a skilled doctor or industrial hygienist so that it can be determined that the patient had absolutely no exposure to asbestos. That means that the patient did not mine or mill asbestos, did not work with a product containing asbestos or did not use a product that may have been contaminated with asbestos. In many of the "background controls" used by other investigators that perform asbestos fiber burden analyses, the history taken usually only states that the patient did not work with an asbestos product. In one case it was documented that the patients did not work with the products but came from an area of the country where there was significant asbestos product manufacturing. That alone should have excluded that population. So when one explores the literature and finds that there is a group of patients exposed to asbestos products and developing disease are being compared to a population where only occupation is the only excluding factor, that is not adequate criteria for calling it a background control. This is referred to as a cohort comparison. One last criterion is that if a patient or the patient tissues are analyzed as background controls and they exhibit either crocidolite or amosite, they should be immediately eliminated as background controls. The reason for them not being considered as background controls is that these fiber types are commercial forms of asbestos that are not found in this country. Therefore, it has to be assumed that the asbestos was from a product containing that type of asbestos and was exposed.

In this author's laboratory, the patients, or the tissues were very critically screened for potential exposure history by very skilled pulmonologists that were trained and worked in coordination with our Environmental Sciences Department. Over the last 35 years this authors laboratory has analyzed tissues from lungs of

207 patients used as background controls. It was only in the initial 25 patients that 3 exposed patients actually slipped through. However, based on finding 1 patient with one amosite fiber, 1 patient with one crocidolite fiber and 1 patient with high concentrations of long chrysotile fibers were the only ones that ultimately proved after extensive further questioning of the family, it determined that these three patients were in fact exposed.

Another criterion to be considered is the timing of background controls, when they were taken compared to the patient that is being analyzed. It has become very apparent that the numbers of asbestos fibers that are being found in patients both exposed and those of background controls have been declining over the years. The phenomenon is the result of the outlawing of most uses of asbestos. Therefore, workers are no longer exposed to asbestos and asbestos products and only those that had been in the past will present with asbestos in their tissues. Another criterion to consider is that over time even the commercial amphiboles will be decreased due to dissolution in the body and removal from the primary site of entrance, presumably the lung. It is a well-known and documented fact that chrysotile has a relatively short half-life in human tissue as compared to amphiboles and therefore, even high exposures of chrysotile, may not be detected in an asbestos fiber burden analysis many years later. It should, however, be noted that chrysotile fibers are not totally removed from the lungs in weeks or months making them relatively non-toxic. Only very long thick fibers are removed from the lung in this period of time. Chrysotile fibers as long as a few hundred micrometers in length can reach the periphery of the lung and once there can be present for years before they are broken down and transported out of the lung or to other tissues. One of the most common hallmarks of a chrysotile exposure is the residual tremolite that one finds in an analysis. Tremolite is a known contaminate of chrysotile that is an amphibole and therefore is more resistant to rapid breakdown and removal. Tremolite tends to be shorter in length and is frequently taken up by macrophages and moves with the smaller broken down chrysotile as compared to the commercial amosite and crocidolite type asbestos. These factors all apply to the background population. Over the 35 plus years of

---

Current levels of asbestos fiber burden observed in digests of lung tissue from our autopsy and surgical population with no history of asbestos exposure. All fibers regardless of size are counted.

[#]Chrysotile type asbestos: Range 0–30,000 fibers/gram wet weight lung
Mean 857 fibers/gram wet weight lung

[*]Amphiboles type asbestos: Range 0–345 fibers/gram wet weight lung
Mean 10 fibers/gram wet weight lung

[+,#]Chrysotile & Amphibole: Range 0–690 fibers/gram wet weight lung
Mean 20 fibers/gram wet weight lung

[**]Asbestos bodies: Range 0–1 bodies/gram wet weight lung
Mean <1 body per gram wet weight lung.

---

[*]*Amphiboles include: tremolite.*

[**]*Asbestos bodies counted by light microscopy of cytocentrifuge preparations. Levels are too low to be detected by electron microscopy.*

[+]*The combination of chrysotile and amphibole fiber burdens represent only cases from the 35 case pool studied where both types of fibers were seen together.*

[#]*100% of the fibers counted were less than 5 μm in length and 100% of those fibers were less than 1 μm in length.*
[@]*All amphiboles fibers were tremolite.*

Table 2.
This table illustrates the range, means and types of asbestos found in the lungs of patients that have had absolutely no exposure to asbestos except for the air they breathe in the New York metropolitan area.

looking at tissue analyses and background controls, it is clear that the amount of background seen is also decreasing. It was once believed that individuals just breathing the air in New York City or for any other city in the world, people would have millions of asbestos fibers in their lungs. This author does not believe that it is true any longer. Based on the most current study group of background controls, it has been determined that no matter how sensitive the testing is done, the great majority of individuals do not exhibited any asbestos in their lungs. The few that have been shown to have asbestos, is restricted to finding very short, less than 1 μm in length, chrysotile fibrils and similarly sized tremolite and nothing else. The results of the analyses of 35 patients meeting all the criteria mentioned above as background controls are shown in **Tables 2–4** for the tissues commonly analyzed in the laboratory.

From a techniques point of view, it is imperative that the analyses of the patient are done with the same degree of sensitivity as the background controls.

Current levels of asbestos fiber burden observed in digests of paratracheal and parabronchial lymph node tissue from our autopsy and surgical population with no history of asbestos exposure.
#Chrysotile type asbestos: Range 0–690 fibers/gram wet weight lymph node.
Mean fibers/gram wet weight lymph node.
*,@Amphiboles type asbestos: Range 0–690 fibers/gram wet weight lung
Mean 20 fibers/gram wet weight lymph node.
+,#,@Chrysotile & Amphibole: Range 0–1380 fibers/gram wet weight lung
Mean 39 fibers/gram wet weight lymph node.
**Asbestos bodies: Range 0–1 bodies/gram wet weight lymph node
Mean <1 body per gram wet weight lymph node.

---

*Amphiboles include: tremolite

**Asbestos bodies counted by light microscopy of cytocentrifuge preparations. Levels are too low to be detected by electron microscopy.

+The combination of chrysotile and amphibole fiber burdens represent only cases from the 35 case pool studied where both types of fibers were seen together.

#100% of the fibers counted were less than 5 μm in length and 100% of those fibers were less than 1 μm in length.
@All amphiboles fibers were tremolite.

Table 3.
This table illustrates the range, means and types of asbestos found in the paratracheal and parabronchial lymph nodes of patients that have had absolutely no exposure to asbestos except for the air they breathe in the New York metropolitan area.

Current levels, 2009–present, of asbestos fiber burden observed in digests of 15 abdominal organs and tissues from our autopsy and surgical population with no history of asbestos exposure. All fibers regardless of size are counted.
Chrysotile type asbestos: Range 0 fibers/gram wet weight abdominal organs and tissues
Mean 0 fibers/gram wet weight abdominal organs and tissues.
*,@Amphiboles type asbestos: Range 0 fibers/gram wet weight abdominal organs and tissues
Mean 0 fibers/gram wet weight abdominal organs and tissues.
Chrysotile & Amphibole: Range 0 fibers/gram wet weight lung
Mean 0 fibers/gram wet weight abdominal organs and tissues.
**Asbestos bodies: Range 0 bodies/gram wet weight abdominal organs and tissues.
Mean <1 body per gram wet weight abdominal organs and tissues.

---

*Amphiboles could include: tremolite or anthophyllite.
**Asbestos bodies counted by light microscopy of cytocentrifuge preparations. Levels are too low to be detected by electron microscopy.

Table 4.
This table shows that in patients with no history to asbestos or talc exposure there was no evidence of asbestos in the abdominal organs including any gynecological organs as the ovaries, uterus, fallopian tubes and cervix.

10. Summary and conclusions

Based on what has been presented above shows that it is clear that there are many possible methods for looking at talcum powders for contaminating asbestos and human tissue for the presence of asbestos, talc and talc contaminants such as aluminum silicates and silica. The difference between these techniques and methods are their sensitivity. The ability to identify these structures go from the least sensitive light microscopic methods using XRD, PLM or PCM to SEM with EDS and then to the most sensitive using a TEM and employing all the analytic methods of EDS and SAED. Sensitivity based on this equipment is based solely on the ability for the instruments to resolve the structures. In most, if not all these methods of looking at the material, sensitivity relies on how one prepares the specimen and how much of the specimen one examines. Therefore, when looking for small fibers or particles that contaminate the talcum powder or the human tissue it is a must, especially when not seen by less sensitive techniques as light microscopy, that the samples have to be examined with an analytic TEM, ATEM and an adequate amount has to be viewed to insure that if the contamination is low or very low, it can still be detected. A perfect comparison is the testing for drugs in blood. If one employs the least sensitive instrument and looking at a relatively tiny sample of blood, small amounts of drugs will not be detected and patients or the addict will not be considered positive when in fact they had taken drugs. Therefore, to identify contaminates in cosmetic talcum powder that will cause disease in humans, one must not only employ the proper instrumentation but also analyze an adequate amount of the talcum powder or human tissue preparation.

Author details

Ronald E. Gordon[1,2,3]

1 Department of Pathology, Icahn School of Medicine at Mount Sinai, New York, NY, USA

2 Core Pathology Electron Microscopy Laboratory, Icahn School of Medicine at Mount Sinai, New York, NY, USA

3 Analytic Asbestos Laboratory, Icahn School of Medicine at Mount Sinai, New York, NY, USA

*Address all correspondence to: ronald.gordon@mountsinai.org

IntechOpen

© 2019 The Author(s). Licensee IntechOpen. This chapter is distributed under the terms of the Creative Commons Attribution License (http://creativecommons.org/licenses/by/3.0), which permits unrestricted use, distribution, and reproduction in any medium, provided the original work is properly cited. (cc) BY

References

[1] Virta RL. Asbestos: Geology, mineralogy, mining, and uses. US Department of the Interior: US Geology Survey. Open-File Report No. 02-149. 2003. pp. 1-28

[2] Takahashi K, Landrigan PJ. The global health dimensions of asbestos and asbestos-related diseases. Annals of Global Health. 2016;**82**:209-213

[3] Langer AM, Ashley R, Baden, Berkley C, Hammond EC, Mackler AD, et al. Identification of asbestos in human tissues. Journal of Occupational Medicine. 1973;**15**:287-295

[4] Gordon RE, Fitzgerald S, Millette J. Asbestos in commercial cosmetic talcum powder as a cause of mesothelioma in women. International Journal of Occupational and Environmental Health. 2014;**20**:318-332

[5] Heller DS, Gordon RE, Katz N. Correlation of asbestos fibers burdens in fallopian tubes and ovarian tissues. Americsn Journal of Obstetrics and Gynecology. 1999;**181**:346-347

[6] Ehrlich H, Gordon RE, Dikman S. Asbestos in colon tissue from occupational exposed workers and general population with colon carcinoma. American Journal of Industrial Medicine. 1991;**19**:629-636

[7] Wu M, Gordon RE, Herbert R, Padill M, Moline J, Mendelson D, et al. Lung disease in world trade center responders exposed to dust and smoke. Environmental Health Perspectives. 2010;**118**:499-504

[8] Cralley LJ, Key MM, Groth DH, Lainhart WS, Ligo RM. Fibrous and mineral content of cosmetic talcum products. American Industrial Hygiene Association Journal. 1968;**29**(4):350-354

[9] Grieger GR. Cover Letter Explanation of Analytical Results, Item MA2270. Westmont, IL: McCrone Associates; 1971

[10] McCrone LB, Shimps RJ. Letter Report of Results—Talc Samples to C. F. Thompson. Westmont, IL: McCrone Associates; 1975. p. 22

[11] McCrone LB. Analysis of Talc By X-ray Diffraction and Polarized Light Microscopy, Under Contract to NIOSH. Westmont, IL: McCrone Associates; 1977

[12] McCrone Associates. Report of Analytical Results, Item MA5500, Talc 1615. Westmont, IL: McCrone Associates; 1977

[13] New York University, Department of Chemistry. Report of Analytical Results. September 1972

[14] Rohl AN, Langer AM. Identification and quantification of asbestos in talc. Environmental Health Perspectives. 1974;**9**:95-109

[15] Rohl AN. Asbestos in talc. Environmental Health Perspectives. 1974;**9**:129-132

[16] Rohl AN, Langer AM. Consumer talcum's and powders: Mineral and chemical characteristics. Journal of Toxicology and Environmental Health. 1976;**2**:255-284

[17] Kremer T, Millette JR. A standard TEM procedure for identification and quantification of asbestiform minerals in talc. The Microscope. 1990;**38**(4):457-468

[18] Nititakis JM, McEwen GN, editors. CTFA Cosmetic Talc J4-1. Washington D.C.: Cosmetic, Toiletry and Fragrance Association, Inc.; 1982. CTFA Compendium Method J 4-1.

Asbestiform amphiboles minerals in cosmetic talc. In: Cosmetic Ingredients Test Methods. Washington D.C.: Cosmetic, Toiletry and Fragrance Association; 1990. pp. 1-6

[19] U.S. Pharmacopeial Convention. Official USP 5/1/09-7/31/09 Monographs: Talc, Absence of Asbestos; 2009

[20] U.S. Environmental Protection Agency. Test Method EPA/600/R-93/116—Method for the Determination of Asbestos in Bulk Building Materials; 1993

[21] Asbestos Hazard Emergency Response Act (AHERA). Appendix A to Subpart E—Interim transmission electron microscopy analytical methods, U.S. EPA, 40 CFR part 763, Asbestos-containing materials in schools, final rule and notice. Federal Register. 1987;**52**(210):41857-41894

[22] American Society for Testing and Materials. ASTM D6281-09, standard test method for airborne asbestos concentration in ambient and indoor atmospheres as determined by transmission electron microscopy direct transfer. West Conshohocken, PA: ASTM International; 2009

[23] American Society for Testing and Materials. ASTM D5755, standard test method for microvacuum sampling and indirect analysis of dust by transmission electron microscopy for asbestos structure number surface loading. West Conshohocken, PA: ASTM International; 2011

[24] American Society for Testing and Materials. ASTM D5756, standard test method for microvacuum sampling and indirect analysis of dust by transmission electron microscopy for asbestos mass surface loading. West Conshohocken, PA: ASTM International; 1998

[25] American Society for Testing and Materials. ASTM D6480, standard test method for wipe sampling of surfaces, indirect preparation, and analysis for asbestos structure number concentration by transmission electron microscopy. West Conshohocken, PA: ASTM International; 1999

[26] International Standards Organization. ISO 10312, Ambient air: Determination of asbestos fibres— Direct-transfer transmission electron microscopy procedure. 1995

[27] International Standards Organization. ISO 13794, Ambient air: Determination of asbestos fibers—Indirect transmission electron microscopy method. 1999

[28] Su S-C. d-Spacing and interfacial angle table for indexing zone-axis patterns of amphibole asbestos minerals obtained by selected area electron diffraction in transmission electron microscope. American Society for Testing and Material (ASTM). 2003-2004:6251-6298

[29] Yamate G, Agarwall SC, Gibbons RD. Methodology for the measurement of airborne asbestos by electron microscopy. EPA Draft Report Contract #68-02-3266; 1984

[30] Wylie AG. Discriminating amphibole cleavage fragments from asbestos: Rationale and methodology. In: Proceedings of the VIIth: International Pneumoconioses Conference: Exposure Assessment and Control Asbestos. 1990. pp. 1065-1069

[31] Wylie AG, Virta RL, Russek E. Characterizing and discriminating airborne amphibole cleavage fragments and amosite fibers: Implications for the NIOSH method. American Industrial Hygiene Association Journal. 1985;**46**(4):197-201

[32] Campbell WJ, Blake RL, Brown LL. Selected Silicate Minerals and Their Asbestiform Varieties, Mineralogical

Definitions and Identification-Characterization. Bureau of Mines Information Circular 8751. Washington D.C.: U.S. Department of the Interior; 1977

[33] Harper M, Lee EG, Slaven JE, Bartley DL. An inter-laboratory study to determine the effectiveness of procedures for discriminating amphibole asbestos fibers from amphibole cleavage fragments in fiber counting by phase-contrast microscopy. Annals of Occupational Hygiene. 2012;**56**(6):645-659

[34] Kelse JW, Thompson CS. The regulatory and mineralogical definitions of asbestos and their impact on amphibole dust analysis. American Industrial Hygiene Association Journal. 1989;**50**:613-622

[35] Miller A, Teirstein AS, Bader ME, Bader RA, Selikoff IJ. Talc pneumoconiosis. Significance of sublight microscopic mineral particles. American Journal of Medicine. 1971;**50**:395-402

[36] Volume 100C. Arsenic, metals, fibres, and dusts in IARC Monographs on the Evaluation of Carcinogenic Risks to Humans. Lyons, France: World Health Organization; 2012. pp. 219-316

[37] Millette JR. Procedure for the analysis of talc for asbestos. The Microscope. 2015;**63**:11-20

[38] Roggli VL, Pratt PC. Chapter 4: Asbestosis. In: Roggli VL, Greenberg SD, Pratt PC, editors. Pathology of Asbestos-Associated Diseases. Boston, Massachusetts: Little Brown & Co; 1992. pp. 77-108

[39] Langer AM, Selikoff IJ, Sastre A. Chrysotile asbestos in the lungs of persons in New York City. Archives of Environmental Health. 1971;**22**:348-361

www.ingramcontent.com/pod-product-compliance
Lightning Source LLC
Chambersburg PA
CBHW081233190326
41458CB00016B/5769